The Politics of Transdisciplinarity

This book discusses collaborative research as both a product of social and epistemic control, and as a process of dealing with it. It offers fresh multi-disciplinary perspectives on old questions that are gaining new urgency with the rise of participatory, transdisciplinary and transformative research.

The volume addresses the complexity of collaborative research at the interface of science, policy and society and sheds light on a common dilemma: researchers and their collaborators tackling issues that require political and knowledge-based control. At the same time, collaborative research that involves diverse publics is difficult to predict or regulate. By examining the interplay of power and knowledge in these collaborations, the book offers insights into how researchers navigate the dilemma of social and epistemic control. This exploration is crucial for understanding the politics of transdisciplinarity. Featuring contributions from various fields including transdisciplinary sustainability studies, science and technology studies, policy analysis, participatory research and more, the book discusses different methodologies, practices, theories and adaptations in response to the control dilemmas inherent in collaborative research to offer a deeper understanding of the nuanced relationship between power and knowledge within collaborative research.

The Politics of Transdisciplinarity will be a key resource for the reflexive researcher working with collaborative approaches at the interface of science, policy and society. It was originally published as a special issue of *Social Epistemology*.

Jeremias Herberg is a sociologist and practitioner at the interface of environmental politics and research. He heads the presidential department of the German Environment Agency and held a visiting professorship at Ruhr University Bochum, Germany. He publishes broadly on social dimensions of sustainability. He co-founded the *Journal of Political Sociology*.

Ulli Vilsmaier is a geographer by training and specialized in inter- and transdisciplinary research and higher education. Her research focuses on designing, accompanying, implementing and evaluating boundary-crossing research and on methods for boundary work and integration. She has significant international experiences with research development and transformations of academic institutions.

The Politics of Transdisciplinarity

Collaboration and Control at the Interface of Science, Policy and Society

Edited by
Jeremias Herberg and Ulli Vilsmaier

Routledge
Taylor & Francis Group
LONDON AND NEW YORK

First published 2025
by Routledge
4 Park Square, Milton Park, Abingdon, Oxon OX14 4RN

and by Routledge
605 Third Avenue, New York, NY 10158

Routledge is an imprint of the Taylor & Francis Group, an informa business

British Library Cataloguing in Publication Data
A catalogue record for this book is available from the British Library

ISBN13: 978-1-032-84968-3 (hbk)
ISBN13: 978-1-032-84970-6 (pbk)
ISBN13: 978-1-003-51586-9 (ebk)

DOI: 10.4324/9781003515869

Typeset in Myriad Pro
by Newgen Publishing UK

Publisher's Note
The publisher accepts responsibility for any inconsistencies that may have arisen during the conversion of this book from journal articles to book chapters, namely the inclusion of journal terminology.

Disclaimer
Every effort has been made to contact copyright holders for their permission to reprint material in this book. The publishers would be grateful to hear from any copyright holder who is not here acknowledged and will undertake to rectify any errors or omissions in future editions of this book.

Contents

Citation Information

The chapters in this book were originally published in *Social Epistemology*, volume 34, issue 4 (2020). When citing this material, please use the original page numbering for each article, as follows:

Introduction

Social and Epistemic Control in Collaborative Research – Reconfiguring the Interplay of Politics and Methodology
Jeremias Herberg and Ulli Vilsmaier
Social Epistemology, volume 34, issue 4 (2020), pp. 309–318

Chapter 1

The Limits of Epistemic Control, the Powers of Actualization, and the Moral Economies of a Fictional Collective
Judith Igelsböck
Social Epistemology, volume 34, issue 4 (2020), pp. 319–329

Chapter 2

Untrol: Post-Truth and the New Normal of Post-Normal Science
Katharine N. Farrell
Social Epistemology, volume 34, issue 4 (2020), pp. 330–345

Chapter 3

Designing a Transformative Epistemology of the Problematic: A Perspective for Transdisciplinary Sustainability Science
Esther Meyer and Daniela Peukert
Social Epistemology, volume 34, issue 4 (2020), pp. 346–356

Chapter 4

Navigating between Complexity and Control in Transdisciplinary Problem Framing: Meaning Making as an Approach to Reflexive Integration
Basil Bornemann and Marius Christen
Social Epistemology, volume 34, issue 4 (2020), pp. 357–369

Chapter 5

Knowledge Decolonization à la *Grounded Theory: Control Juggling in Research Situations*
Maria De Eguia Huerta
Social Epistemology, volume 34, issue 4 (2020), pp. 370–381

Chapter 6

Identity Politics: Participatory Research and Its Challenges Related to Social and Epistemic Control
Stefan Böschen, Martine Legris, Simon Pfersdorf and Bernd Carsten Stahl
Social Epistemology, volume 34, issue 4 (2020), pp. 382–394

Chapter 7

The Phase Zero: Why Collaborative Research Is Not Co-Designed but Scripted
Jeremias Herberg
Social Epistemology, volume 34, issue 4 (2020), pp. 395–407

Notes on Contributors

Basil Bornemann is a senior researcher and lecturer at the Department of Social Sciences, University of Basel, Switzerland, and the Center for Religion, Economy and Politics, University of Zürich, Switzerland. His research interests include governance for sustainability, with a particular focus on the state; sustainability transformation conflicts and democracy with a focus on food and energy; and principles and practices of transformative sustainability research. He has an interdisciplinary background in environmental sciences, a doctoral degree in political science, and a venia legendi in political science and sustainability research.

Stefan Böschen is Professor for "Society and Technology" at the Human Technology Center at RWTH Aachen, Germany. He is also the Director of the Käte Hamburger Center "Cultures of Research" and Spokesperson for the Human Technology Center at RWTH. He studied chemical engineering, philosophy and sociology, and holds a diploma in chemical engineering and a doctoral degree and habilitation in sociology.

Marius Christen is a lecturer at the Sustainability Research Group, University of Basel, and the Kompetenzzentrum Public Management, University of Bern, Switzerland. A philosopher by training, he researches on sustainability theory, sustainability governance and inter- and transdisciplinarity. His actual transdisciplinary research focuses on sustainability governance in Swiss Cantons. Additionally, he consults governments and public administrations in implementing sustainability.

Maria De Eguia Huerta is the Director of Lighthouse Global, developing learn and research transformative methodologies in collaboration with small innovative initiatives in three continents working on social and ecological sustainability. She holds a communication and cultural studies degree, a gender and diversity postgraduate degree and a PhD in politics. She has collaborated as a consultant, researcher and trainer with universities, development and cooperation agencies, NGOs and local governments on issues such as of multidimensional human development, postdevelopment, intersectionality, knowledge production and global North–South power relations.

Katharine N. Farrell is an ecological economist, member of Pi Sigma Alpha, and holds degrees in political science, urban policy analysis and management, environmental engineering and a PhD in politics. She is currently Visiting Professor with the Interdisciplinary Center for Development Studies of the Universidad de los Andes, Colombia, after having successfully defended her Habilitation with the Agricultural Economics Department of the Humboldt-Universität zu Berlin in 2018, where she served 2010–2016 as Assistant Professor in Resource Economics. She has worked as a senior researcher with the Danish National Environmental Research Institute (NERI), University of Aarhus, Denmark, L'Institut de Ciència i Tecnologia Ambientals (ICTA), Universitat Autònoma de Barcelona, Spain and the Helmholtz Centre for Environmental Research (UFZ) in Leipzig, Germany and was awarded a Marie Curie Individual Driven Fellowship to conduct the research project Accountability and Legitimacy of Governance Institutions that support Viable Environments (ALIVE) at UFZ from 2006 to 2008.

Jeremias Herberg is a sociologist and practitioner at the interface of environmental politics and research. He heads the presidential department of the German Environment Agency and held a visiting professorship at Ruhr University Bochum, Germany. He publishes broadly on social dimensions of sustainability. He co-founded the *Journal of Political Sociology*.

Judith Igelsböck is a social researcher of emerging technologies and innovation. She has worked in various research areas, including human–computer and human–robot interaction, work studies, science and technology studies, and innovation and organization studies. Judith approaches her research projects in a post-disciplinary and experimental mood and enjoys working with artists. One of her current obsessions is the speculative fusion of digital twin discourse and twin fiction. You are welcome to reach out to Judith via email (judith.igelsboeck@tum.de).

Martine Legris is Research Engineer at Lille University, Centre for Administrative, Political and Social Studies and Research (CERAPS), France. She studied sociology and contemporary history. She holds a PhD in sociology and a master's degree in history. She is an internationally recognized scholar in sociology, has published in Social and Human Sciences (SHS) and is active across the fields of participatory sciences and participatory democracy, and critical research.

Esther Meyer is a lecturer in sustainability communication at Leuphana University, Lüneburg. In her position at the "Stiftung Deutsche Fernsehlotterie", she advises and supports non-profit projects in the funding area of intersectional and participatory development of quarters and neighbourhood. With her collectively managed social start-up, she supports organizations in communicating their social and ecological impact from a perspective of cooperation, transformative learning, and education for sustainability. She studied Philosophy & Economics at the University of Bayreuth, holds a doctorate from the Faculty of Sustainability at Leuphana University and has completed training in systemic organizational development.

Daniela Peukert is Professor of Design for Sustainability Transformation at Tomorrow University of Applied Sciences and Program Director of the Master's program "Sustainability, Entrepreneurship and Technology" in cooperation with Vienna University of Economics and Business. With a background in product design and sustainability sciences, she works on sustainable products that have a positive impact. She completed her PhD at Leuphana University Lüneburg and is an expert in transdisciplinary design methods and the role of design in sustainability transformations. In her research, she explores the epistemic qualities of prototyping practices and material metaphors to foster integration and knowledge co-production within transdisciplinary processes and works towards a planet-centric design approach.

Simon Pfersdorf is a seasoned researcher with a background in political science and a doctorate in sociology. Throughout his career, he has focused on exploring societal participation in the production of scientific knowledge. Currently, as the CEO of CONTRACT GmbH he serves as a dedicated coach and consultant, guiding organizations and executives through transformative changes.

Bernd Carsten Stahl is Professor of Critical Research in Technology at the School of Computer Science of the University of Nottingham. His interests cover philosophical issues arising from the intersections of business, technology and information. This includes ethical questions of current and emerging of ICTs, critical approaches to information systems and issues related to responsible research and innovation.

Ulli Vilsmaier is a geographer by training and specialized in inter- and transdisciplinary research and higher education. Her research focuses on designing, accompanying, implementing and evaluating boundary-crossing research and on methods for boundary work and integration. She has significant international experiences with research development and transformations of academic institutions.

Introduction: Social and Epistemic Control in Collaborative Research – Reconfiguring the Interplay of Politics and Methodology

Jeremias Herberg and Ulli Vilsmaier

ABSTRACT

In this article we argue that the notion of control poses a critical conceptual and historical connection between scientific and political power. While many meanings of control originate in the sciences, concepts of experimentation, care, and learning currently translate into increasingly decentralized governance concepts, be it through market-logics or surveillance technologies. That is, epistemic and social control is co-constituted. Collaborative research plays a transformative but paradoxical role in this interplay: Science and Technology Studies scholars have leveraged powerful critiques against techno-scientific control and have shaped practical modes of transdisciplinary research. However, the critique of techno-scientific control is increasingly mixed up with post-truth controversies, and the appeal to inter- and transdisciplinary collaboration has been appropriated by neoliberal science policy. The historical conundrum culminates in a practical dilemma: Collaborative researchers seek to overcome the very regimes of techno-scientific control that the sciences are bound to co-produce. Can they shift the control regimes that they are part of? Collaborative research requires a critical and pragmatic standpoint with regard to both the methods of politics and the politics of methodology. This book seeks to come to terms with the inherent contradictions of collaborative research and make useful proposals with regard to its political potential.

> Now examine the reasons by which control is exercised in the lifeboat scenario: The two leaders are armed, let's say, with .38 revolvers – twelve shots and eight potential opponents. They can take turns sleeping. However, they must still exercise care not to let the eight rowers know that they intend to kill them when land is sighted. Even in this primitive situation force is supplemented with deception and persuasion. The leaders will disembark at point A, leaving the other sufficient food to reach point B, they explain. They have the compass and they are contributing their navigational skills. In short they will endeavor to convince the others that this is a cooperative enterprise in which they are all working for the same goal. [...]
>
> William Burroughs, The Limits of Control (1978)

According to the Beat poet William Burroughs, the exercise of control is never complete but relies on the collaboration of the controlled. The controlled, in turn, find themselves in a schizophrenic position: they ignore and simultaneously accept their subordinate role. To change the social imbalance, they would have to be very self-aware and very careful not to overturn the lifeboat.

In this book, we discuss a theme that involves, but exceeds, the socio-psychological imprint of control. Various studies from Science and Technology Studies (STS), Environmental Studies, Postcolonial Theories, and Interpretative Policy Analysis have pointed out the deep entwinement between control in society on the one hand, and control as applied in the scientific study of nature and society, on the other. Inter- and transdisciplinary approaches have been proposed in response to the reinforcing power-play between scientific expertise and political power. However, research *on* and experience *with* collaborative research suggests that one cannot easily evade the involvement in social and epistemic control. Be it neoliberal funding schemes or the misuse of academic authority – the most difficult contexts often motivate, or at least shape collaborative research. Amidst the climate crisis or post-truth politics, it is very possible that collaborative researchers find themselves stuck in a true lifeboat situation. How do control settings squeeze the collaborative room to maneuver, they may ask. How, in turn, can collaborative endeavors recognize or even challenge a given status quo?

Our starting point is that the entwinement of epistemic and social control works very delicately in collaborative processes and delimits the capacity to overcome undesirable power relations. We analyze and explore ways to reconfigure the relationship between control and collaboration from within and seek to come to terms with the inherent contradictions and political potential of collaborative research. The critical context against which the contributions should be read is delineated in the following. This context is conceptual and historical.

Why Reconfigure Social and Epistemic Control?

Control has various meanings. As a verb it can, for instance, imply to monitor or to regulate someone or something; it can involve direct and indirect influence; it can take place from near or far away; and it can relate to a process or a product. The most common reading is the notion of political rule and social surveillance: think of the unravelling of social control in surveillance studies or in the Chicago School of urban sociology, or of Hobbes' Leviathan as a symbol of statehood and regulatory rationality. Many readers – hopefully intrigued to see a special issue on control – will expect a considerable share of political criticism.

Yet, to question control settings goes beyond the position of distant scrutiny. Any observer, when building a cognitive relationship to an object of study, exercises a restraining authority on their own gaze and surroundings. Therefore, especially when criticizing political control regimes, one is well advised to question one's own exercise of epistemic control (cf. Bourdieu 1992). In another meaning, as an intransitive verb, control implies to control *for* something. This aspect, while often perceived as a paternalistic gesture, involves a benevolent oversight or even a moment of care. When controlling a disease, for instance, one seeks to reduce the incidence or severity of an event. This is not necessarily a detached – but instead a deeply emotional – relationship to the object of attention. Entire fields of study, be they public health or climate science, have arguably endorsed this precautionary meaning of control in the political realm. Similar to Burroughs' lifeboat scenario, climate scientists and climate diplomats care about existential threats. Therefore they simultaneously seek to leverage and offset the governing power of industrialized nations. Can scientists find methods of counteracting a control regime from within? Where are the tricksters of counter-control, in scientific disciplines or beyond? The paradigmatic method for a fine-tuned sensorium of control in the sciences is the experiment. Experimental control 'purifies' the complex sociotechnical context and allows for an isolated relationship between subject and object (Latour 1993). To conduct an experiment, however, involves constant negotiation with the controlled web of instruments, hypotheses, and settings that eventually facilitate a narrow set of inferences (Rheinberger 2001). Only rarely, '[…] scientific practice becomes effectively frozen at moments of interactive stabilization […]' (Pickering 1995, 102). Control is thus not unlimited either in Burroughs' lifeboat scenario or experiments. It can hardly be generalized, but is best seen as a contradictory and imperfect practice.

To complicate the matter, the sciences do not offer a straightforward toolkit of counter-control. Certainly, there are social skills that help to institutionally improve, or to individually adapt to control regimes (cf. Lyall 2019). Yet, many critical takes on the role of science in society have highlighted an entwinement in which methods of research always interact with methods of governance, be they regulatory, coercive, or caring. The discerning gaze of the scientist and the logic of the experiment undergirds modern methods of spatial planning (Scott 1998), was instituted in the history of management and administration (Beniger 2009), and is reproduced through education (Freire [1996] 1970). Less directly, it has also translated into political forms of continuous experimentation such as computerized surveillance, in which '[P]ower can no longer be summarized by that totalitarian symbol of centralized command and control.' (Zuboff 2015, 82). Finally, it is important to note that control is fundamentally anchored in the ethical underpinnings of industrial societies. Merchant (1981), along with many others, is concerned with the way that control is implicated in the liberal male subject, as she points out with regard to Boyle's notion of experiment: 'The new image of nature as a female to be controlled and dissected through experiment legitimated the exploitation of natural resources.' (Merchant [1981] 1989, 189) The shrewd affect of control is indirectly related to destructive effects in nature and society. Hence, to critically engage with a control setting is a practical and intellectual endeavor; it goes beyond political criticism and reaches into scientific self-reflection.

It is not far-fetched to say that STS, Gender Studies, and Environmental Studies have been motivated by the critique of both regulatory and scientific forms of control. Here and in other interdisciplinary fields, we can encounter more or less cunning tricksters of counter-control. The entwinement of both social and epistemic control implies, however, that control is difficult to reconfigure, as most famously pointed out by Michel Foucault: 'The fundamental codes of a culture – those governing its language, its schemas of perception, its exchanges, its techniques, its values, the hierarchy of its practices – establish for every man, from the very first, the empirical orders with which he will be dealing and within which he will be at home.' (Foucault 2005, xxii). The difficulty of escaping one's embeddedness in control regimes, as symbolized by Burroughs' lifeboat scenario, can therefore not be overcome without recognizing both epistemic and social dimensions. Socially, one can certainly adopt the 'soft voice' of the intellectually flexible and adaptive interdisciplinary researcher. Or one may join the 'loud voices' that open-up research programs by means of science policy or institutional leadership (cf. Lyall 2019). Yet, these proposals fall short in questioning the way that collaborative research is epistemically embedded in control regimes. Corresponding proposals can be seen in the social sciences. Even where the connectedness of the researcher and their research subject is most obviously taken into consideration, however, it is the aim in most qualitative and quantitative research to minimize the researcher's influence. This methodical form of control has only been challenged at the margins of the scientific landscape, such as in ethnography and ethnomethodology where the reflexive relation to one's own positionality and associated role and identity of the researcher, and their relationship with the field is systematically integrated in the research process.

Many interdisciplinary responses fall short in acknowledging the inescapable need to reconfigure control in and through the research collaboration. This book therefore engages in an endeavor of what Steve Fuller has called 'modal power.' That is, we seek to circumscribe and shape the possibility space that is left for collaborative research endeavors amidst prevailing control regimes (Fuller 2018). Similar to the lifeboat scenario, one needs to consider both situated imbalances and a general predicament: First, it is at the fringes of social and epistemic sectors or disciplines that we can best reflect upon the conditions, epistemologies, theories, or practices that facilitate social rank and epistemic authority. In these interstices, where conventional rules lose their binding effect, the co-creation of emancipatory research collectives is most likely to succeed (Herberg 2018; Vilsmaier, Brandner, and Engbers 2017). Second, the role of the sciences must be accepted as a manifestation of the history of control regimes. Otherwise, a critique of control may even reproduce some of the positivist and technocratic implications. That means in practice that the connection between scientific and political methods must be critically explored before projects of collaborative research can even begin to change undesirable power relations. The following section gives a historical account of why this dual strategy – the practice within and the critique of control settings – is necessary to transform socio-ecological crises or other lifeboat scenarios.

The Rise of the Collaboration-and-Control Conundrum

Can collaboration reconfigure control? We presented the authors of this issue with the challenge of reflecting on the affects and effects of control while grappling with reconfigurative approaches in their empirical context. This conundrum has historically emerged throughout a genealogy that we can only sketch briefly. As political economists, historians of science, and control theoreticians have pointed out: today's control regimes, which are predominantly based on market-logics and surveillance technologies, have shaped how scientists learn and collaborate.

In the course of the wars of the 20th century, military capacities led to the rise of computer-driven regimes of political and technoscientific control (Agar 2003). In the course of computerization and other technocratic policies, the role of scientific institutions was increasingly rationalized as economic input (Popp Berman 2012), which has placed science policy squarely on the frontlines of neoliberal regimes (Mirowski 2002). Many fields of scientific research, for instance economics (ibid.), but also countercultural practices of knowing and organizing (Turner 2010), were increasingly wrapped up in mechanistic and computer-like metaphors of control. Especially when related to military purposes and to international relations, cross-disciplinary collaboration and a clear orientation towards politically defined problems became broadly accepted values among scholars and policy makers. In the course of a reconciliatory international diplomacy from the 1970s onwards, collaborative forms of research that span non-academic domains formed as part of a generic appreciation of epistemic and social complexity (Leendertz 2018; Nowotny 2005). Well-funded think tanks such as the International Institute for Applied Systems Analysis in Austria expanded methodologies such as operations research or complex modelling to the interdisciplinary sciences of global change (Schrickel 2017). Climate science, for example, came to focus on problems of control in complex environments (Edwards 2010). As a result of this genealogy, many collaborative research traditions, for instance in the environmental arena, engaged in both the epistemic refinement and the critical reflection of political governance. This dual strategy illustrates an ambiguous relationship with social and epistemic control.

How to politically leverage collaborative research so that it can reconfigure control regimes? Science policy itself became more dispersed in the 1970s, especially in the form of patents or quantitative evaluation criteria (Popp Berman 2012). These trends did not go uncommented, but critical scholarship accompanied that history, and has (re-)activated alternative modes of collaboration (Turner 2013). In a less optimistic fashion, poststructuralist thinkers have developed theories that highlight the decentralized distribution and the inescapability of social and epistemic forms of control (cf. Gilbert and Goffey 2015; cf. Igelsböck this issue). The core of Foucault's contribution shows that 'there is always only one episteme that defines the conditions of possibility of all knowledge, whether expressed in a theory or silently invested in a practice (2005, 183).' Largely meant as emancipatory projects, control theories, as exemplified by Deleuze's famous Postscript on the Societies of Control (1992), have reinvigorated critical engagement with practical forms of knowledge in psychiatry, education, and planning. More recently, Gilbert and Goffey even conclude their discussion of control theories with a call for democratic experimentation (2015).

This brings us to the question of whether the very practices of collaborative research can and should be subject to value-oriented reform. The democratic experimentation with epistemic and social control can in itself look back on a long history. Since the 1970s – the same decade that brought about complex tools of governance – emancipatory alternatives have emerged that challenge techno-scientific imaginaries of control. In Latin America, participatory action research was developed. Inspired by the theological and pedagogical liberation movements (Freire [1970] 1996; Fals Borda 2001), collaborative research was established in-between civil society and academics, thereby introducing a heterogeneity of ways of knowing and acting as a means for transforming unjust and imbalanced societal conditions. Similar to attempts initiated by dissenters throughout the history of modern science, such as Saul Alinsky in 1930s Chicago (1989), these approaches seek to distribute power in defining research objectives and mechanisms of legitimization. Interestingly, concerns for justice were also applied to questions of method. The question of 'Whose voice is heard?' raised by G. C. Spivak (1988) carries not only a social-political meaning; it likewise incorporates an epistemic dimension concerning what *can* be said and heard. The opposition to political methods sparked a new politics of method. The response to control was combined with epistemic reconfigurations.

Another exemplary approach has developed in the environmental arena in Europe and the US. The overwhelming insights into the destruction of sustainable futures prompted scholars to question the power and appropriateness of traditional modes of research. Informed by insights from STS, so-called transdisciplinary approaches have been conceptualized and put into practice, that take into consideration the multiplicity of perspectives on given phenomena (e.g. Nowotny, Scott, and Gibbons 2001; Funtowicz and Ravetz 1993). The aim of these approaches was to create research spaces that are accessible for people with less academic authority and social privilege. This inclination, which goes beyond a mere egalitarian desire, is historically connected to the epistemic appreciation of complexity since the 1970s. As Helga Nowotny et al. stated, linear control is to be replaced by a more collective and cautious approach to complex problems: 'The old belief in (technocratic) predictability may be being eroded because of the growing recognition that the world is a non-linear place and full of surprises, but the belief in the efficacy of other forms of anticipations and in the self-fulfilling prophecy of expectations is on the increase.' (2001, 39). While the linearity of techno-scientific control was subject to critical scrutiny, those early collaborative forms of research were upheld as a reflexive form of controlling for uncertainty. Similar arguments are characteristic of the school of post-normal science (Funtowicz and Ravetz 1993; cf. Farrell in this issue). This and other transdisciplinary approaches, however, have been criticized as inconsequential. The notion of epistemic control, which basically assumes that researchers can neatly isolate complex research subjects, persists despite the acknowledgement of increasing social complexity (Wesselink and Hoppe 2011; Shinn 2002). Despite or possibly because of this inconsequential standpoint, the policy impact of transdisciplinary schools of thought was substantial. For instance in the US and the Netherlands environmental policy makers have widely embraced a post-normal concept of the science–policy interplay (Steel et al. 2004; Hoppe 2009). This rise of transdisciplinary research was criticized as a Trojan Horse for industry-oriented science policy: 'Now the tide has turned. Against the backdrop of tightening research budgets and four decades of almost uninhibited growth, the orientation of science to societal values is seen as a new virtue, a promising source of legitimation.' (Weingart 1997, 611). Transdisciplinary research has become part of the policy-making toolbox.

From this perspective both sides of the collaboration-and-control conundrum – the increasing reflexivity about control regimes and the increasing acknowledgement that researchers need to collaborate – are best discussed as an intertwined process. While collaborative endeavors often come across as an emancipatory response to control, they are historically shaped by the predicament to simultaneously address and escape the power structures of academia and politics. Collaborative responses to the collaboration-and-control conundrum are always contradictory approximations. Being the somewhat academic version of Burroughs' life-boat scenario, the unwanted involvement in techno-economic control regimes has impelled diverging pathways of collaborative research. Some seek to carve out a research practice from in-between institutionalized fields that are grounded in difference (Vilsmaier, Brandner, and Engbers 2017) while others struggle for a counter-hegemony against the epistemological and ontological regime of techno-science (Escobar 2018). Many promising responses – including those gathered in this book – do not prioritize a particular way of producing knowledge, but instead explore social, political, and epistemic conditions, spaces and temporalities, and methodological responses. This becomes particularly relevant where research crosses not only disciplinary boundaries but the boundaries of science itself, thereby creating research spaces that span different ways of knowing, acting, and being (Vilsmaier, Brandner, and Engbers 2017).

We even argue, as an interim conclusion, that the historical conundrum has increasingly culminated in a dilemma. Collaborative researchers seek to overcome the very regimes of techno-scientific control that the sciences have helped to co-produce. While there is a necessity to respond, all responses are necessarily unsatisfactory: On the one hand, one can endorse the trickiness of the lifeboat situation and thereby try to overcome undesired power relations from within. The tricksters of political counter-control would engage in an opportunistic, yet subversive craft. However, collaborative research has so far not superseded but rather complicated decentralized regimes of control. On the other hand, one may attempt to find alternative rules, criteria, or methods as part of more emancipatory approaches. For this perspective on epistemic reform, it seems necessary to accept, however, that control remains an all-pervading dimension of collaborative research. Reconfiguring control thus involves reflecting, reconnecting, and constantly balancing the social and epistemic facets of research. If successful, collaborative research approaches, for instance the newly emerging approaches of transdisciplinary research can – ever so slightly – shift the existing forms of authority and privilege. Nonetheless, they are very likely to co-produce some of the established or newly altered forms of control.

The most current rendition of this dilemma – we may now even call it a battle – concerns scientific facts. The notion that facts may be countered has become popularized through reference to 'alternative facts' or even falsehoods. Composed of technological conditions of possibility, automation, political opportunism, mistrust, fear, narcissism, and arbitrariness, the battlefield spans an ever-deeper belief in the scientific rationales and associated methods of knowledge production, machines, social media, powerful political and economic agents, and a revived notion of the population as demographic masses. In this journal and other venues, scholars have re-assessed their role in the history of decentralizing epistemic authority. Some came to self-critical conclusions with regard to the rise of post-truth politics (Latour 2004). Steve Fuller, by contrast, re-affirms the notion of symmetry, which was established in STS to study the creation of both facts and non-facts, and sees post-truth politics as a public form of endorsement (Fuller 2016), while Sergio Sismondo replied that 'epistemic democratization has to involve more equitable political economies of knowledge' (2017, 3). As a result of the many post-truth controversies, collaborative researchers of various strands search for a political role that helps to somewhat democratize the definition and solution of publicly relevant problems.

The value of these struggles is that the historicity of fact-claims has become a graspable feature of public political life. In this context, the methodologies of inter- and transdisciplinary research may also be challenged in their political innocence: How do institutional and discursive control regimes inhibit or enable collaborative processes? Is collaborative research a way to reconfigure social and epistemic control? How can the entanglement of social and epistemic control be made transparent and incorporated into research methodologies?

The Contributions

In convening a broad range of responses to the collaboration-and-control conundrum, this book discusses collaborative research as both a product and a process of social and epistemic control. In focusing on collaborative research, the presented theme encompasses not only questions of control in knowledge production and research, but also tackles the relationship between epistemology and politics, cognition and practice, social roles and epistemic ambitions: Firstly, the current relationship between positivist research and populist politics delimits the countervailing role and legitimacy of collaborative research. Judith Igelsböck and Katharine N. Farrell critically elaborate on political epistemology and propose epistemic reconfigurations. Secondly, epistemic and social control comes to the fore in the predetermination of knowing itself. The contributions by Esther Meyer and Daniela Peukert, and by Basil Bornemann and Marius Christen, discuss how the understanding of a problem delimits the way we frame and approach phenomena and formulate research questions. Thirdly, epistemic and social control shapes the conditions of research, i.e. the political, financial, and institutional settings. The contributions by Maria De Eugia Huerta, Stefan Böschen, and Jeremias Herberg discuss how such settings include and exclude, prioritize or marginalize, certain types and subjects of research.

Inspired by Jim Jarmusch's movie 'The Limits of Control' (2009), Judith Igelsböck theorizes the processes of winning/losing/loosening epistemic control and conducts a thought experiment that follows the quest to reconfigure epistemic control in collaborative research. She holds that even in the post-truth climate of skepticism about the authority of science, the ever inaccessible 'dream of epistemic democracy' remains desirable. Igelsböck therefore seeks to identify 'loopholes' in the circular forms of epistemic control and relies on Daston's conception of a 'moral economy of science' (1995) to fictionalize a collective of social researchers of 'non-control.' Researchers in this collective are equipped to 'cut-up' pre-recorded research realities that too narrowly define ways of knowing and their modes of circulation.

In her theoretical essay, Katharine N. Farrell unravels the tragic interaction between the increasing responsibility that characterizes the Anthropocene and the unresponsiveness of post-truth politics. She argues that the fetishizing use of truth claims, as currently practiced by the US government, is a continuation of the ways in which the positivist and technological values of the European Enlightenment are broadly appropriated, while social justice ideas were left abandoned. In reference to critical theory and the approach of post-normal science, she shows that populist forms of epistemic control further undermine the control over and wise governance of environmental crises. To counter this dynamic that she fittingly describes as the politics of 'untrol,' Farrell endorses Terry Eagleton's notion of 'hope without optimism' ([2005] 2009). A politically grounded understanding of epistemological complexity (Farrell [2005] 2009) can enable institutions, for instance the Intergovernmental Science-Policy Platform on Biodiversity and Ecosystem Services, to address the 'how-to-questions of the Anthropocene.'

Esther Meyer and Daniela Peukert depart from a notion of problems as used in sustainability studies, and elaborate an epistemological reconfiguration. The connection between the notion of problems and the socio-epistemic features of collaborative research has not been discussed or realized as a methodological basis so far, although problems represent the epistemic and political focus of sustainability studies. Meyer and Peukert suggest that a critical take on problems and control can be opened-up by learning from the poststructuralist notion of 'The Problématique' and from design as a field of thinking and doing. They create a reconfigurative dialogue across both these areas by focusing on the differentiation of research subject and object. Their proposal of a 'thinking practice of problematic designing' entails a process that deliberately interrelates epistemic conditions on the one hand and designed drafts that allow for manifold transformative dimensions on the other. This approach preserves a procedural character while engaging with the politically defined and severe problems that transdisciplinary sustainability scholars focus on.

Basil Bornemann and Marius Christen trace similar tendencies to deal with sustainability problems, but propose a different interpretation of how problems take shape in collaborative research. Starting from the observation that some approaches of transdisciplinary research lead to a leveling of problem perceptions, Bornemann and Christen ask how a greater diversity can be maintained. They argue for a reflexive form of integration of perspectives in which the polar tension between complexity and control may not be overcome, but can be collectively interpreted in a situated practice of 'meaning making.' From this perspective, 'theoretical structuring,' 'grasping diversity,' 'debating complexity,' and 'framing the problem' are consecutive steps towards a form of collaborative research that achieves a balance between complexity-induced paralysis on the one hand, and the perils of social and epistemic control of research on the other. The authors illustrate and reflect this proposal by means of a research project on sustainability governance in Swiss cantons.

Maria De Eguia Huerta's autoethnographic approach addresses the 'politics of methodology' from within a specific collaborative process. Her scientific and practically situated involvement as a young urban Spanish woman in rural Bolivia confronted her with what she calls 'ethico-political choices.' With a dialogue between the empirical strategy of grounded theory on the one hand and with the writing of postcolonial theorists on the other, De Eguia Huerta grapples with the practicability of epistemic justice. While illustrating that ethical certainty with regard to control is unattainable, she showcases playful practices that enable a momentary sharing of control. To introduce a 'decolonizing momentum' into her research amidst Bolivian and German communities, she advocates for a practice of 'control juggling.' Intimate artifacts such as personal photographs, she suggests, can help researchers and practitioners to articulate and to mediate a situated reflection on roles.

Stefan Böschen, Martine Legris, Simon Pfersdorf, and Bernd Carsten Stahl use several case studies out of a larger research project to discuss the dynamic of 'identity politics' between civil society organizations and researchers. They point out that most collaborative research is organizationally configured within short-term projects while simultaneously being bound to a multiplicity of communities. The observed collaborative practices are highly influenced by socio-epistemic tactics used to control the epistemic hybrid forum, while preserving previously held identities. Böschen et al. distinguish between weakly tied identity politics and strongly tied identity politics, claiming that strongly tied identity politics leaves only limited space for situational maneuver and re-adjustment in collaborative research.

Herberg's qualitative case study reconstructs the formation of a research project in which he was involved. On this basis he introduces the notion of 'scripting control,' which expands the analytical concept of scripts to also entail a conception of temporal power dynamics. By contrast to an understanding of control as established by systems analysts since the 1960s and as sustained in recent sustainability studies, he claims that project initiators do not backcast or forecast the current means against a set target state, but rather impose an interwoven set of desires, resources, and expectations. While collaboration is often imagined as a democratic exercise, the momentary synchronization of interests is overrun by the rule of time. Collaborative research processes, which he likens to Paulo Virilio's notion of 'dromocracy' (Virilio [1977] 2006), can therefore be better studied and evaluated from a diachronic perspective.

The contributions all show that control is key to the epistemic quality and normative legitimacy of knowledge production. They propose various epistemic concepts, practices, or approaches to reconfigure the dynamics of social and epistemic control. It is our hope that readers will appreciate this collection as being simultaneously a critical reading *of* and a daring engagement *with* the inescapability of control in collaborative research practice.

Disclosure statement

No potential conflict of interest was reported by the authors.

ORCID

Jeremias Herberg http://orcid.org/0000-0003-4388-796X

References

Agar, J. 2003. *The Government Machine. A Revolutionary History of the Computer*. Cambridge, MA: MIT Press.
Alinsky, S. 1989. *Rules for Radicals. A Practical Primer for Realistic Radicals*. New York: Vintage Books.
Beniger, J. 2009. *The Control Revolution: Technological and Economic Origins of the Information Society*. Cambridge, MA: Harvard university press.
Bourdieu, P. 1992. "The Practice of Reflexive Sociology (The Paris Workshop)." In *Pierre Bourdieu & Loïc Wacquant: An Invitation to Reflexive Sociology*, 216–260. Chicago: University of Chicago Press.
Burroughs, W. S. 1978. "The Limits of Control Semiotext(e)." *Schizo-Culture* 2 (2): 38–42.
Daston, L. 1995. "The Moral Economy of Science." *Osiris* 10: 2–24. doi:10.1086/368740.
Deleuze, G. 1992. "Postscript on the Societies of Control." *October* 59, 3–7.
Edwards, P. N. 2010. *A Vast Machine: Computer Models, Climate Data, and the Politics of Global Warming*. Cambridge, MA: MIT Press.
Escobar, A. 2018. *Designing for the Pluriverse. Radical Independence, Autonomy and the Making of Worlds*. Mexico City: Duke University Press.
Fals Borda, O. 2001. "Participatory (Action) Research in Social Theory: Origins and Challenges." In *Handbook of Action Research*, edited by P. Reason and H. Bradbury, 27–37. London: Sage.
Farrell, K. N. [2005] 2009. *Making Good Decisions Well: A Theory of Collective Ecological Management*. Aachen: Shaker Verlag.
Foucault, M. 2005. *The Order of Things*. London, NY: Routledge.
Freire, P. [1996] 1970. *Pedagogy of the Oppressed*. (et al.). London: Penguin Books.
Fuller, S. 2016. "Embrace the Inner Fox: Post-truth as the STS Symmetry Principle Universalized." Social Epistemology Review and Reply Collective. Accessed 16 October 2019. https://social-epistemology.com/2016/12/25/embrace-the-inner-fox-post-truth-as-the-sts-symmetry-principle-universalized-steve-fuller/#comments
Fuller, S. 2018. *Post-truth: Knowledge as a Power Game*. Vol. 1. London: Anthem Press.
Funtowicz, S. O., and J. R. Ravetz. 1993. "Science for the Post-Normal Age." *Futures* 25: 739–755. doi:10.1016/0016-3287(93)90022-L.
Gilbert, J., and A. Goffey. 2015. "Control Societies: Notes for an Introduction." *New Formations: A Journal of Culture/Theory/Politics* 84 (84): 5–19. doi:10.3898/NEWF:84/85.INTRODUCTION.2015.
Herberg, J. 2018. "Public Administrators in the Europeanization of Risk Governance–Co-Creation Amidst the Political Heterarchy. Culture." *Practice and Europeanization* 3 (2): 42–58.
Hoppe, R. 2009. "Scientific Advice and Public Policy: Expert Advisers' and Policymakers' Discourses on Boundary Work." *Poiesis & Praxis* 6 (3–4): 235–263.
Jarmusch, J. 2009. "Archival Talks: Jim Jarmusch, 'The Limits of Control'." Film at Lincoln Center. *YouTube*. Accessed June 2019. https://www.youtube.com/watch?v=v62aGp0v64k
Latour, B. 1993. *We Have Never Been Modern*. Cambridge, MA: Harvard University Press.
Latour, B. 2004. "Why Has Critique Run Out of Steam? from Matters of Fact to Matters of Concern." *Critical Inquiry* 30 (2): 225–248. doi:10.1086/421123.
Leendertz, A. 2018. "Towards a New Diplomatic History of Transatlantic Relations." In *The TransAtlantic Reconsidered*, edited by C. A. Lerg, S. Lachenicht, and M. Kimmage, 120–138. Manchester: Manchester University Press. Accessed 11 October 2019. https://www.manchesterhive.com/view/9781526119384/9781526119384.00011.x
Lyall, C. 2019. *Being an Interdisciplinary Academic: How Institutions Shape University Careers*. Cham: Springer.
Merchant, C. [1981] 1989. *The Death of Nature: Women, Ecology, and the Scientific Revolution*. New York: Harper.
Mirowski, P. 2002. *Machine Dreams: Economics Becomes a Cyborg Science*. Cambridge: Cambridge University Press.
Nowotny, H. 2005. "The Increase of Complexity and Its Reduction: Emergent Interfaces between the Natural Sciences, Humanities and Social Sciences." *Theory, Culture & Society* 22 (5): 15–31. doi:10.1177/0263276405057189.
Nowotny, H., P. Scott, and M. Gibbons. 2001. *Re-thinking Science: Knowledge and the Public in an Age of Uncertainty*. Cambridge, UK: Polity Press.
Pickering, A. 1995. *The Mangle of Practice: Time, Agency, and Science*. Chicago: University of Chicago Press.
Popp Berman, E. 2012. *Creating the Market University: How Academic Science Became an Economic Engine*. Princeton: Princeton University Press.
Rheinberger, H.-J. 2001. *Experimentalsysteme und epistemische Dinge. Eine Geschichte der Proteinsynthese im Reagenzglas*. Frankfurt: Suhrkamp.
Schrickel, I. 2017. "Control versus Complexity: Approaches to the Carbon Dioxide Problem at IIASA." *Berichte zur Wissenschaftsgeschichte* 40 (2): 140–159. doi:10.1002/bewi.201701821.

Scott, J. C. 1998. *Seeing Like a State: How Certain Schemes to Improve the Human Condition Have Failed*. New Haven, CT: Yale University Press.
Shinn, T. 2002. "The Triple Helix and New Production of Knowledge." *Social Studies of Science* 32 (4): 599–614.
Sismondo, S. 2017. Post-truth? *Social Studies of Science* 47 (1): 3–6. doi:10.1177/0306312717692076.
Spivak, G. C. 1988. "Can the Subaltern Speak?" In *Marxism and the Interpretation of Culture*, edited by C. Nelson and L. Grossberg, 271–313. Chicago: University of Illinois Press.
Steel, B., P. List, D. Lach, and B. Shindler. 2004. "The Role of Scientists in the Environmental Policy Process: A Case Study from the American West." *Environmental Science & Policy* 7 (1): 1–13. doi:10.1016/j.envsci.2003.10.004.
Turner, F. 2010. *From Counterculture to Cyberculture: Stewart Brand, the Whole Earth Network, and the Rise of Digital Utopianism*. Chicago: University of Chicago Press.
Turner, F. 2013. *The Democratic Surround: Multimedia and American Liberalism from World War II to the Psychedelic Sixties*. Chicago: University of Chicago Press.
Vilsmaier, U., V. Brandner, and M. Engbers. 2017. "Research In-between: The Constitutive Role of Cultural Differences in Transdisciplinarity." *The Transdisciplinary Journal of Engineering & Science* 8: 169–179. doi:10.22545/2017/00093.
Virilio, P. [1977] 2006. *Speed and Politics*. Cambridge: The MIT Press.
Weingart, P. 1997. "From 'Finalization' to 'Mode 2': Old Wine in New Bottles?" *Social Science Information* 36: 591–613. doi:10.1177/053901897036004002.
Wesselink, A., and R. Hoppe. 2011. "If Post-normal Science is the Solution, What is the Problem?: The Politics of Activist Environmental Science." *Science, Technology, & Human Values* 36 (3): 389–412.
Zuboff, S. 2015. "Big Other: Surveillance Capitalism and the Prospects of an Information Civilization." *Journal of Information Technology* 30 (1): 75–89. doi:10.1057/jit.2015.5.

The Limits of Epistemic Control, the Powers of Actualization, and the Moral Economies of a Fictional Collective

Judith Igelsböck

ABSTRACT

This essay narrates from a collective of social scientists giving up on the phantasy of 'being in,' or 'having' epistemic control, not – however – on the 'dream of epistemic democracy'. This community does not feel 'pre post-truth' nostalgia. And when there is a special issue asking 'what are possible reconfigurations of collaborative research beyond control?', they reply: 'we are not sure, but we are in the mood to figure it out.' This mood is not related to the naïve assumption that knowledge production was not a powerful control machine in its own right, or that issues of control could be ignored or dismissed as vanities. It is built upon the feeling that dynamics of epistemic control cannot be escaped, but can and should be played and experimented with. The essay makes use of Lorraine Daston's conception of a 'moral economy of science' to fictionalize the 'mental state' of such a collective of social scientists and the 'emotional forces' integral to their ways of performing research.

Introduction

> I was told over and over in my life by authority figures: This is not how the world works. And I am so fed up with hearing this because the world just works the way you, your consciousness, can perceive it. Imagination is the most powerful thing we have. (Jarmusch 2009)

In research, 'epistemic control' refers to techniques to rationalize, systematize, and routinize the ways in which we know, including the identification of legitimate contributors to knowledge production and dissemination processes, as much as proper forms of knowing, instruments, procedures etc. Epistemic control has often been debated as an issue of 'having' and 'being in,' or 'not having' and 'losing' control. The latter is tangible in current debates about the roles and responsibilities of the (social) sciences in the so-called post-truth era, for instance. At the same time, there are also discussions about the need to widen the scope of sources of knowledge and expertise in scientific knowledge production – far beyond what is currently considered to be 'scientific'. Such calls for 'loosening' epistemic control are often related to narratives of a complexification of societal dynamics and the urge to find new ways of dealing with them. This paper will lay out debates revolving around (a contemporary loss of, or the need to loosen) epistemic control. It will then argue that these often seem to come with recommendations about how to deal with epistemic control issues, while not thinking in theoretical terms on the practices of winning, losing, loosening, or sharing epistemic control in the first place. The paper mobilizes control theories to address this gap. Having conceptualized epistemic control, the paper fictionalizes a collective of scientists who – building on this concept of epistemic control – do not necessarily strive for 'being in' or 'having' epistemic control; who do not give up on the dream of 'epistemic democracy'[1], however. The essay uses Lorraine Daston's concept of a 'moral economy of science' (1995) for drawing together the hopes, feelings, and beliefs of such a community and their practices of epistemic 'non-control', which is not to be confused with 'irresponsible' research, as will be stressed.

Debating Epistemic Control and the Dream of Epistemic Democracy

In the last years, the issue of epistemic control has been connected to a variety of debates and concerns: battles for resources between disciplines, the paradoxes characterizing 'knowledge societies', the public (mis-)understanding of science and knowledge production processes, the rationalization of scientific knowing, or the complexity of contemporary societal challenges. Within these debates, we can roughly distinguish between two movements.

One can be described as a call for 'loosening epistemic control' and striving towards an 'epistemic democracy' in which several ways of knowing and experiencing are valued, heard, and taken into consideration (see e.g. Maasen and Peter 2005; Epstein 1996). 'Transdisciplinary knowledge production' or 'extended peer review' (see e.g. Gibbons and Nowotny 2001; Funtowicz 2001) can be mentioned as institutional answers to such calls. In the search of solutions for societal problems with utmost complexity, such as anthropogenic climate change, transdisciplinarity is invoked as an attempt to systematically include expert knowledge from outside of the science system into scientific knowledge production processes at all stages (see e.g. Thompson-Klein 2004). Felt et al. (2016) have stressed in their empirical study of transdisciplinary research in the field of environmental sustainability that involving a broader variety of actors who might have something to contribute to the understanding and solution of a problem – as is envisaged in transdisciplinary research – is not as trivial as it might seem. This is not necessarily the case because researchers lack ideas about who these actors could be and how to enroll them in knowledge production processes (even though many of the analyzed cases had stayed with a classical linear model of knowledge production in which the researchers were responsible for problem definition and knowledge production, while additional actors were involved towards the end of a project for the dissemination of their results beyond the classical science communities). Rather, one major difficulty that keeps hindering the broader uptake of ideas of opening up knowledge production processes could lie in the fact that current scientific regimes of valuation did not correspond to an epistemic lifestyle that keeps reaching out to the various ways in which our worlds can and could be known outside of the boundaries of the respective disciplines. So, the important question to ask might not be whether researchers should give away epistemic control, but whether they are granted the freedom to do this without being dismissed from their scientific communities and career tracks.

In a less 'institutional' version of the 'dream of epistemic democracy', Law and Ruppert (2013) welcome and embrace 'modes of knowing' that are usually dismissed in research. They point out, 'there are many ways of knowing and knowing well. And academic styles of doing so might be revitalized by looking behind their current boundaries' (19). As a facet of this attempt, affect studies and non-representational (or more-than-representational) theories call for paying attention 'to forces – visceral forces beneath, alongside, or generally other than conscious knowing, vital forces insisting beyond emotion – that can serve to drive us towards movement, towards thought and extension' (Gregg and Seigworth 2010, 1). This movement can be read as critique of instrumental reason and 'the textual' as preferred techniques used to know scientifically.

The second movement of epistemic control is an attempt to regain and fight for control rather than loosening it. Joan Fujimura (1998) is interpreting the issue of epistemic control as a battle for resources between scientific disciplines, by pointing out that these battles were not solely intellectual ones but had stakes reaching far beyond the careers of individual academics. Huutoniemi (2016) presents interdisciplinary quality control as a 'counter-force' to disciplinary control and autonomy. 'Like scholarly critique and response in general, a more critical attitude between disciplines is likely to improve the reliability of knowledge', Huutoniemi (2016, 179) stresses. She discusses the topic with an eye on what has been debated as 'paradoxes of expertise and authority' (see e.g. Bijker, Bal, and Hendriks 2009) in 'knowledge societies' (see e.g. Stehr 2001a). Contemporary 'knowledge societies' do not seem to be characterized by a growing pool of certainties we can build on. Several notions have analytically put into focus 'how we do not and not yet know' and hint to an increasing concern for what '[...] lies outside of regular and quantifiable events' (Gross 2010, 2). The mobilization of knowledge as a 'resource for action' (Gross 2010, 51), moreover provokes a tendency of putting knowledge and expertise into question, a disposition to 'continuous revision' of what can count as knowledge and expertise, and subsequently an accelerated rhythm of regulation and deregulation. Related to that, Stehr (2001b) diagnoses a 'fragility of modern societies'. Weingart (2001) describes an increasing coupling of science and politics, which challenges the public image of science as operating on the values of impartiality and objectivity. Related dynamics have been pointed out by Bijker and co-authors as the 'paradox of scientific authority':

> The cases in which scientific advice is asked most urgently are those in which the authority of science is questioned most thoroughly. (Bijker, Bal, and Hendriks 2009, 1)

Against the backdrop of these struggles, social scientists, and specifically scholars of science and technology studies (STS), have put into question their ways of portraying scientific processes of knowledge production (see e.g. Latour 2004; Collins 2009). This is not utterly surprising, as it was one of their key achievements to demystify scientific knowledge production by highlighting the constructed nature of facts, or the contingencies and resistances accompanying highly controlled knowledge production processes. Of course, it might never have been their intention to challenge the trust in science when deconstructing the 'Snow White fairy-tale of scientific objectivity' (Farrell 2011) by 'insisting on the rhetorical nature of truth', or by 'mak[ing] clear that official ideologies about objectivity and scientific methods are particularly bad guides to how scientific knowledge is actually made' (Haraway 1988, 576). As Latour claims, '[t]he question was never to get away from facts but closer to them, not fighting empiricism but, on the contrary, renewing empiricism' (Latour 2004, 234). Some scholars put effort into 'repair work' highlighting that 'constructed' was not to be confused with 'random' and 'unreal' or also 'dishonest', 'fraudulent', or 'unprofessional' (see e.g. Hacking 1999).

In the wake of the so-called post truth era, the control over knowledge production and circulation seems increasingly 'out of hand'. This has effects on the way in which epistemic control is debated these days. It is more or less common sense that the sciences fight a battle for epistemic control against the 'merchants of doubt' as Oreskes and Conway (2010) call them. 'How can researchers respond to organized, sophisticated and persistent attempts to undermine science?' Oreskes and Conway ask. How to react in a situation in which 'merchants of doubt' (such as representatives of the tobacco industry or 'climate skeptics') 'create a whole scientific Potemkin village' (687)? STS scholar Michael Lynch (2017) expresses worries about undesirable uses of STS insights on how to manufacture factual knowledge:

> Science studies opened up the cognitive terrain to those concerned to enhance the impact of democratic politics on science, but in doing so, it opened that terrain for all forms of politics, including populism and that of the radical right wing. (597)

Collins, Evans, and Weinel (2017) fear that further attempts at 'democratizing' science would 'invite exactly the skepticism about experts and other elites that now dominate political debate in the US and elsewhere', as Lynch (2017, 596) points out. Fuller (2016), suggests we 'embrace *our* responsibility for the post-truth world and call forth our vulpine spirit to do something unexpectedly creative with it' (3). The latter is certainly an inspiring call. But what could 'doing something creative' mean in this regard?

The above mentioned scholars often seem to utter their opinions and concerns around the issue of how to deal with problems revolving around epistemic control in a post-truth world. They do not seem to elaborate more extensively on how to theorize epistemic control in the first place though. How does one arrive in a state in which one can 'exercise' epistemic control? What level of epistemic control can be reached? What are the processes of gaining and losing epistemic control? The following part is devoted to providing a possible answer to these questions by performing a thought experiment mobilizing theories of control. In a subsequent effort, the paper fictionalizes the moral economies of a collective of social scientists and their take on epistemic 'non-control' based on these thoughts.

The Society of Epistemic Control – A Thought Experiment

Let's get started with our thought experiment on epistemic control with this image of a rather gloomy stage setting of Henrik Ibsen's play 'An Enemy of the People (En Folkefiende)' as written in 1882, and performed at the Residenztheater in Munich 2019.[2] The play revolves around questions of how to establish truths, but this should be secondary, we mostly work with the stage setting (see Figure 1). A glass house. No doors for stepping out of it. Not on the front and also not on the back side. Narrow corridors. A few rooms to meet and discuss inside. The glass house is moving slowly. Spinning, rotating about itself, to be more precise. Sometimes to the left and sometimes to the right. Sometimes it stops. Everything that is happening, happens inside of the glass house.

In our thought experiment, the glass house is the 'society of epistemic control': the space in which we know. And while we aim to gain control over what we know and believe, and strive to earn credibility for it, we are learning that we are being controlled by the realities we find (such as registers of trustworthy forms of knowledge and processes in which these can be established). We cannot but disagree with the play's protagonist 'Dr. Stockmann' who desperately claims that the strongest human in the world is whoever stands most alone. In the society of epistemic control, we are dependent on a range of allies to achieve an outcome in which we have stabilized our position in a way, so we can get close to the feeling that we 'have' epistemic control. And we can never seem to rest, because the glasshouse keeps spinning and turning. Whoever strives for epistemic control, needs to invest in 'credibility', and ride up and down the *cycle of credit* (Latour and Woolgar 1986)[3]: articles, arguments, data equipment, money, grant, recognition, grant, money, equipment, data, arguments, articles ..., etc. Who strives for sharing epistemic control with who isn't perceived as a legitimate knowledge producer, risks to lose credibility. The path towards epistemic control seems to be narrowly predefined. Look at the hallway in the image! It is pretty restrictive; claustrophobic even. And like its idol – the society of control[4] – the society of

Figure 1. Stage setting: En Folkefiende by Henrik Ibsen as performed at the Residenztheater in Munich in 2018.

epistemic control is characterized by 'limitless postponements'. Epistemic control can only be had in the future.

> [...] [T]he receipt of reward is just one small portion of a large cycle of credibility investment. [...] The essential feature of this cycle is the gain of credibility which enables reinvestment and the further gain of credibility. Consequently, there is no ultimate objective to scientific investment other than the continual redeployment of accumulated resources. (Latour and Woolgar 1986, 197–98)

In the attempt to gain or loosen epistemic control, we find ourselves controlled by a 'pre-recorded' future reality (see below); A future reality of historically grown (dis-) trust in distinct forms of knowledge, institutions, and epistemic practices; A future reality that privileges certain actors while excluding others (what Dotson (2014) would call 'epistemic oppression'), and predefines the conditions of what and how we can know.

The Powers of Actualization

How 'real' are these future realities? And what does their realness tell us about possible reconfigurations of research 'beyond' control? Is there any escape from our epistemic control society? Searching for answers provided by 'control'-scholars, such as Foucault (1977) and Burroughs (1978), we cannot develop much hope for acting outside of our society of (epistemic) control. But there might be other ways. Let's draw on a paradox outlined in a similar way on 'power' and 'control' by different authors. First, the 'power' argument by Latour (1986):

> The problem of power may be encapsulated in the following paradox: when you simply have power – in potential – nothing happens and you are powerless; and when you exert power – in actu – others are performing the action and not you. (264)

In 'the limits of control' Burroughs (1978) makes a similar argument:

> All control systems try to make control as tight as possible, but at the same time, if they succeeded completely, there would be nothing left to control. [...] When there is no more opposition, control becomes a meaningless proposition. (Burroughs 1978, 40)

Burroughs identifies time – needed to exercise control – as the 'basic impasse' of all control machines. Tying this back to the pre-recorded reality of the cycle of credit, both its power and weakness resides within the fact that it is 'real' (in the sense of potentially well-rehearsed and (relatively) stabilized), as much as it needs actualization for its continuous existence.[5]

How to challenge the epistemic control society when being controlled by it? We can borrow from Burroughs who develops the 'cut-up method' for his fictional agent to 'challenge pre-recorded reality' in a fiction in which 'there is no time or real "reality", in which reality – indeed "flesh reality" – itself is represented as a film' (see Ayers 1993, 228). Analogous to Burroughs' cut-up technique, we can identify our resistant and transformative agency precisely in the moments in which we are needed to prevent the 'actual' from 'falling from the plane [of immanence] like a fruit' (Deleuze [1977] 2002, 150; see also footnote 5). The epistemic control society controls us, but also needs us to rearticulate and actualize pre-recorded realities. And in this very moment lies the resistant power.

The following section introduces a fictional collective of researchers – the social scientists of non-control – who are not only aware of the reality of epistemic control societies, but also of the powers of actualization. How they are intending to use it, can be read in the subsequent part.

The Moral Economies of a Fictional Collective

This part fictionalizes on the social scientists of 'non-control'. It fictionalizes – because it wants to be an ode to the powers of actualization while not denying the strength of what we have come to call 'pre-recorded realities' with Burroughs. It fictionalizes even though many of the sources it draws on are stemming from actual social scientists, so that the collective might be more 'facticious'[6] than purely 'fictional' (whatever that would be). But who knows whether the mobilized actors would want to be considered members of this collective, or have somebody – who does not know them – talking about their mental states. So, fictionalizing is a literary disclaiming technique as much as a philosophical practice.

Why choose 'moral economies' (Daston 1995) when describing the fictional collective of social scientists of 'non-control'? Moral economies refer to 'psychology at the level of whole cultures, or at least subcultures, one that takes roots within and is shaped by quite particular historical circumstances' (5), to the 'interior of Merton's black box' (7).

> In contrast to Mertonian norms, moral economies are historically created, modified, destroyed, enforced by culture rather than nature and therefore both mutable and violable; and integral to scientific ways of knowing. (Daston 1995, 7)

Moral economies are bound to the everyday practices of researchers, to the choices they make with regard to research subjects and methods, to the sources they find inspiring and trustworthy, to the standards they apply to research processes. Importantly, moral economies cannot be reduced to strategic choices of researchers. Moral economies enact 'a psychology that goes beyond the calculus of self-interest, strategically deployed to the ends of disciplines – or career-building.' (24) Instead, the focus lies more on 'how intellectual work is saturated with moral, emotional and aesthetic elements' (24). The work with moral economies bring the collective's mental states together with micro-practices of their research. It does not exclude emotions and feelings from their everyday epistemic practice, and does not assume that emotions and values are purely individual, outside from or distorting science. 'They are science', as Lorraine Daston highlights (6).

No 'Pre Post-Truth' Nostalgia

The fictional collective of social scientists of non-control does not share pre-post truth nostalgia. It does not wish to go back in time, in which the authority of science was widely unquestioned, and factual knowledge was more likely to be respected and accepted as such by its audience (if this ever was the case). Not because it wants everyday life to be extra challenging, but because it does not see a point in 're-establish[ing] a hierarchy between knowledge and its presumed opposite, non-knowledge, or antiknowledge', as Marres (2018, 423) brings to the point. 'No, we don't want our facts back', the collective agrees with Marres. This would mean giving up on the dream of epistemic democracy in favor of the authority of science. And while epistemic democracy might be an ideal that can never be reached, the collective strives to continuously learn from and experiment with different ways of knowing, seeing, feeling, and being in the world (see e.g. Law 2016). It acknowledges 'knowledge democracy [as] a reconstructive project [that] requires transformation of epistemic ideals, including that of facticity' (Marres 2018, 441).

In the Mood for Experimentation

'What took us so long?', the fictional collective of social scientists of non-control asks. 'Hasn't all of this been there for a while?'

> Along with Haraway and Latour, we will argue that science intervenes in nature and politics, and that this approach provides a much better way of understanding what scientific activity is than do old notions about how science discovers and described reality. (Asdal, Brenna, and Moser 2007, 9)

Asdal, Brenna, and Moser (2007, 9) describe 'the politics of intervention' as one of the key constructivist STS 'lessons'. Hacking made a similar point a while ago: With our scientific apparatuses, we create the phenomena we seek to observe, discover, understand and portray (see Hacking 1983, 220ff). For a long time, however, it seemed that we haven't been up for exploring the potentials these lessons hold for our own ways of performing (collaborative) research. We kept playing describers, observers, interpreters of the reality-making of others, and saw our vocation in the analysis of other scientists' politics of intervention.

'Why has critique run out of steam' (Latour 2004) is, surely, one of the most prominent, but also maybe most abstract examples of calls for investing in the exploration of techniques of (re-) assembling (instead of mostly debunking) realities (and this includes the realities in which we create knowledge). More recently, a number of authors have been re-evaluating experiments as a form of social research, not necessarily the ones of the positivist kind (see e.g. Blackman 2014; Lezaun, Muniesa, and Vikkelsø 2013; Pickering 2002). Pickering (2002) sees the perfect spokespersons for a turn from the representational idiom to the performative one in three cyberneticians.

> While [classical scientists] seek to pin the world down in timeless representations, cybernetics directly thematizes the unpredictable liveliness of the world, and processes of open-ended becoming. While classical science has thus been an epistemological project aimed explicitly at knowledge production, cybernetics is an ontological project, aimed variously at displaying, grasping, controlling, exploiting and exploring the liveliness of the world. (430)

And Pickering also motivates the fictional collective to play with 'pre-recorded realities'. 'Like most scholarly authors, I have grown accustomed to think that representation is what we do. [...] [B]ut such need not exclusively have to be the case' (ibid). He adds that '[o]ne can, of course, still be interested in epistemology and knowledge production, but my conclusion was that one should see scientific knowledge as constitutively bound up with the dance of human and nonhuman agency, as I romantically labelled it, rather than as a self-contained topic for enquiry in itself.' (Pickering 2002, 431.)

Lezaun, Muniesa, and Vikkelsø (2013) flirt with experimental practices as carried out by Moreno, Lewin, Bion, Milgram, and Zimbardo in post-world war II. They worked in a way in which they provoked realities, 'activated the latent energies to everyday life and in doing so reveal reality' (279). Inspired by 'the experimental principle of the natural sciences, namely that the degree of realism of any representation is dependent of the sophistication of the apparatus of experimental intervention', these experimental scholars aimed to 'generate alternative forms of conviviality' (280). While critics once dismissed these experiments as 'imperfect method[s] to re-present social phenomena' (280), Lezaun et al. highlight their capacity to 'produce vivid and otherwise unavailable renderings of social reality [...] in a particularly demonstrative and explicit form' (280).

Choosing to keep up the dream of epistemic democracy, and striving for the exploration of collaborative forms of research, the fictional collective of 'non-control' gets inspired by scholars who are 'putting aside the classic repertoire of representational validity' (290), and experiment with the cutting-up of pre-recorded realities; all of them operating at the interface of the virtual and the actual, the fictional and the factual. This can mean 'cultivat[ing] a sense of the possible that concerns, but does not owe its existence to the ways in which the actual determines the distribution of what is probable' (Wilkie, Savransky, and Rosengarten 2017, 7). This can also be 'going beyond observation and description' for the performance of 'imaginary economics', and 'adopting a strategy of imitation, magnification, simulation and parody', in a mode of knowledge creation that 'is that of an essayist not that of an analyst' (Cameron 2014, 113). Another such practice has been labelled as 'fictocriticism', a 'hybrid writing technique' that – according to Kerr and Nettlebeck (1998, 3) – 'moves between the poles of fiction ("invention"/'speculation') and criticism ('deduction'/'explication'), of subjectivity ('interiority') and objectivity ('exteriority').' In a touching outcry about disciplinary control, Carl Rhodes (2015) promotes the technique of 'fictocriticism' for organization studies as a form of contestation of the conservatism of their own discipline and the destructive effects its continuous reproduction has on the work and life of its scholars:

> Our work and the life it involves suffer at the hand of its own conservatism. This spectre that haunts us is a future where 'succeeding generations can continue to make progress, by keeping the science in social science' (Donaldson 2005, 1085). This is a science that coaxes and cajoles into hypotheses, argument, evidence, proof and logic in a dispassionate pursuit that classifies, ossifies and putrefies the terror and wonder of the world desperate to render it knowable and predicable. What then are the possibilities for writing in and from the

> academy in a manner that might somehow allow the heart's instincts to be followed and the vast possibilities of expression to be explored and enjoyed? (Rhodes 2015, 290)

Inspired by all of these ways of dealing with pre-recorded research realities, the fictional collective of non-control opts for experimental cut-ups. In contrast to the prevalent meaning of experimentation in 'modern science', the experimental cut-up is not a controlled arrangement to verify or falsify hypotheses. The collective leaves aside this historically more recent interpretation of the experiment, but reactivates what Mersch (2015) has framed as the 'original meaning' of the 'experimental' alongside of the verbs 'experire/experiens/experio' (experiencing/trying something out/putting something at stake) as well as 'expetere' (happening/occurring) (see 56). This notion of the experimental is one in which the experiment cannot be possessed by the experimenters, in which the experimenting is about experiencing what is exposed, what does not conform, what cannot be calculated. It is thus an ode to defiance and rebellion.

Inside

> 'The "we" is not a sum of "I" 's, but a novelty produced by legacies, concessions, withdrawals, resignations, of the "I." The "we" is less a set of "I's" than the set of the sets of its transmissions.' (Serres 1981, 228)

Last, but not least, the fictional collective of non-control includes themselves in the processes of experimentation. It does not assume that it could experiment and play without being played and experimented with. This is different to the account of epistemic control as developed by Caniglia et al. (2017). They have put effort into coming up with a typology of experiments and the sort of epistemic control the researchers 'have' on the interventions and manipulations in their respective form of experimentation (e.g. lab and field experiments, adaptive experiments, real-world experiments, etc.). In doing so, they are reducing control to the type of knowledge researchers can come up with, while not addressing how the researchers themselves and their epistemologies are subject of experimentation.

Not 'being in', or 'having' epistemic control, but 'playing' with it, and also embracing 'being played' with, is not to be equalized with irresponsibility for the research that is pursued and the realities that are brought into being. Rather, it comes with the certainty that whatever 'methods for knowing and handling the world' are deployed, they have their own 'social life' (Law and Ruppert 2013). They 'establish' and 'format relations', but they are also formatted by the nets in which they are actualized. Most definitely, they are 'used opportunistically by social actors in the systematic pursuit of political, economic and cultural advantage.' (239) Manipulating and knowing the world are not rendered mutually exclusive. In this sense, caring for epistemic democracy, and developing and manipulating in the sense of 'plan[ing], schem[ing], trick[ing], contriv[ing], invent[ing]' (229) is not necessarily a contradiction. It is not, because it comes with passion for learning and 'becoming with' who/what is studied, instead of 'keeping the science in social science' (see above).

Discussion

The article takes up the question of potential reconfigurations of collaborative research beyond control, as posed by the editors of this book. It reviews two current *movements* with regard to epistemic control. One is striving towards 'epistemic democracy' (in a more and less institutional manner), the *countermovement* is arguing for restoring the authority of science in society by setting clear hierarchies between scientific and other forms of knowing. The paper argues that in order to understand what it could mean to go beyond control, the processes of winning/losing/loosening control need to be theorized in the first place. When thinking about research control in theoretical terms – it seems that we can only gain control when also letting ourselves being controlled in a rather tight corset of 'pre-recorded realities' which prescribe – in our case – how knowledge is to be produced. These 'pre-recorded realities' have been grasped as the 'society of epistemic control' in which the knowledge producer is enrolled as credibility investor who is everything else but free when it comes to perform (collaborative) research. We thus might have to remain skeptical about the possibility of going 'beyond control' when thinking about (collaborative) research in its diverse shapes. However, the paper also identifies moments of resistance, moments that allow for the 'cut-up' of prerecorded research realities: the processes of actualization (of the pre-recorded reality). The fictional collective of non-control is betting on the powers in these processes. Instead of developing 'pre post-truth' nostalgia, it develops lust for experimentation and playing with pre-recorded research realities; Their experimentation is not focused on fighting back epistemic control, but on giving it away. Despite dystopian narratives about knowledge production and circulation in the post-truth era, the fictional collective of non-control does not give up on the dream of 'epistemic democracy', and – with an eye on the effects on communal lives – continuously keeps striving to challenge the epistemic nets of habituality with their often unnecessarily restrictive and rationalized accounts of what can be regarded as valid and valuable forms of knowledge and modes of knowing.

The limits of epistemic control in this fiction refer to both the limits to how we can be controlled by the 'society of epistemic control' and the limits to how we can control epistemic processes. Certainly, as compared to many other people and forms of knowing, researchers and their rationalities already find themselves in a privileged position. Instead of putting a lot of effort into actualizing this privilege, the collective of non-control aims to create situations in which new knowledge collaborations can become possible.

Notes

1. Speaking of 'epistemic democracy' as a dream evokes an understanding of democracy as an ideal that cannot be reached, or that cannot be reached once and for all. As Dewey puts it, '[democracy] is an ideal in the only intelligible sense of an ideal: namely, the tendency and movement of something which exists carried to its final limit, viewed as completed, perfected. Since things do not attain such fulfillment but are in actuality distracted and interfered with, democracy in this sense is not a fact and never will be' (Dewey 1927, 148). And yet there isn't anything else to do than put constant effort into the creation of possible realizations of 'epistemic democracy' in the sense of staying attentive to the effects our ways of (non-)controlling, organizing, and restricting knowledge production processes show on our communal lives and vice versa.
2. https://www.residenztheater.de/inszenierung/ein-volksfeind accessed on 15 January 2019.
3. 'The notion of credibility allows the sociologist to relate external factors to internal factors and vice versa. The same notion of credibility can be applied to scientists' investment strategies, to epistemological theories, to the scientific reward system, and to scientific education. Credibility thus allows the sociologist to move without difficulty between these different aspects of social relations in science.' (Latour and Woolgar 1986, 188).
4. Is it possible to draw analogies between Deleuze's interpretation of Foucault's Societies of Control (Deleuze 1992; Foucault 1977), and our epistemic control societies? '[I]n the societies of control one is never finished with anything – the corporation, the educational system, the armed services being metastable states coexisting in one and the same modulation, like a universal system of deformation.' (Deleuze 1992, 4).
5. Deleuze ([1977] 2002) writes about the process of actualization in a chapter on 'the actual and the virtual': 'Philosophy is the theory of multiplicities, each of which is composed of actual and virtual elements. [...] The plane of immanence includes both the virtual and its actualization, simultaneously, without there, being any assignable limit between the two. The actual is the complement or the product, the object of actualization, which has nothing but the virtual as its subject. Actualization belongs to the virtual. The actualization of the virtual is singularity whereas the actual itself is individuality constituted. The actual falls from the plane like a fruit, whilst the actualization relates it back to the plane as if to that which turns the object back into the subject.' (148–149).
6. Facticious: 'artificial, visibly made up, but made up in such a fashion that it carries with it an irreducible element of reality; in other words, the situation cannot be dismissed as simply fictitious.'(Lezaun, Muniesa, and Vikkelsø 2013, 279).

Acknowledgments

I would like to thank the MCTS (Munich Center of Technology in Society) of the Technical University of Munich, the LIT (Linz Institute of Technology) of the Johannes Kepler University and the STS department of the University of Vienna for providing space and time for me to use my imagination.

Disclosure statement

No potential conflict of interest was reported by the author.

References

Asdal, K., B. Brenna, and I. Moser. 2007. *Technoscience: The Politics of Interventions*. Oslo: Oslo Academic Press.

Ayers, D. 1993. "The Long Last Goodbye: Control and Resistance in the Work of William Burroughs." *Journal of American Studies* 27 (2): 223–236. doi:10.1017/S0021875800031546.

Bijker, W., R. Bal, and R. Hendriks. 2009. *The Paradox of Scientific Authority: The Role of Scientific Advice in Democracies*. Cambridge: MIT Press.

Blackman, L. 2014. "Affect and Automaticy: Towards an Analytics of Experimentation." *Subjectivity* 7 (4): 362–384. doi:10.1057/sub.2014.19.

Burroughs, W. 1978. "The Limits of Control." *Semiotext(e): Schizo-Culture* 3 (2): 38–42.

Cameron, A. W. 2014. "Modelling 'Headless': Finance, Performance Art, and Paradoxy." *Journal of Critical Globalisaton Studies* 7: 102–121. http://hdl.handle.net/2381/29021

Caniglia, G., N. Schäpke, D. J. Lang, D. J. Abson, C. Luederitz, A. Wiek, M. D. Laubichler, F. Gralla, and H. Von Wehred. 2017. "Experiments and Evidence in Sustainability Science: A Typology." *Journal of Cleaner Production* 169: 39–47. doi:10.1016/j.jclepro.2017.05.164.

Collins, H. 2009. "We Cannot Live by Scepticism Alone." *Nature* 458 (7234): 30. doi:10.1038/458030a.

Collins, H., R. Evans, and M. Weinel. 2017. "STS as Science or Politics?" *Social Studies of Science* 47 (4): 580–586. doi:10.1177/0306312717710131.

Daston, L. 1995. "The Moral Economy of Science." *Osiris* 10: 2–24. doi:10.1086/368740.

Deleuze, G. 1992. "Postscript on the Societies of Control." *October* 59: 3–7. https://www.jstor.org/stable/778828

Deleuze, G. [1977] 2002. *Dialogues II*. New York: Columbia University Press.

Dewey, J. 1927. *The Public and Its Problems*. New York: Holt.

Donaldson, L. 2005. "Vita Contemplativa: Following the Scientific Method: How I Became a Committed Functionalist and Positivist." *Organization Studies* 26 (7): 1071–1088. doi:10.1177/0170840605053542.

Dotson, K. 2014. "Conceptualizing Epistemic Oppression." *Social Epistemology* 28 (2): 115–138. doi:10.1080/02691728.2013.782585.

Epstein, S. 1996. *Impure Science: AIDS, Activism, and the Politics of Knowledge*. Berkeley: University of California Press.

Farrell, K. 2011. "Snow White and the Wicked Problems of the West: A Look at the Lines between Empirical Description and Normative Prescription." *Science, Technology & Human Values* 36 (3): 334–361. doi:10.1177/0162243910385796.

Felt, U., J. Igelsböck, A. Schikowitz, and T. Völker. 2016. "Transdisciplinary Sustainability Research in Practice: Between Imaginaries of Collective Experimentation and Entrenched Academic Value Orders." *Science, Technology, & Human Values* 41 (4): 732–761. doi:10.1177/0162243915626989.

Foucault, M. 1977. *Discipline and Punish: The Birth of the Prison*. New York: Pantheon.

Fujimura, J. H. 1998. "Authorizing Knowledge in Science and Anthropology." *American Anthropologist* 100: 347–360. doi:10.1525/aa.1998.100.2.347.

Fuller, S. 2016. "Embrace the Inner Fox: Post-truth as the STS Symmetry Principle Universalized." *Social Epistemology Review and Reply Collective*. http://wp.me/p1Bfg0-3nx

Funtowicz, S. 2001. "Peer Review and Quality Control." In *International Encyclopaedia of the Social and Behavioural Sciences*, edited by N. J. Smelser, and P. B. Baltes, 11179–11183. Oxford: Elsevier.

Gibbons, M., and H. Nowotny. 2001. "The Potential of Transdisciplinarity." In *Transdisciplinarity: Joint Problem Solving among Science, Technology, and Society*, edited by J. T. Klein, W. Grossenbacher-Mansuy, R. Häberli, A. Bill, R. W. Scholz, and M. Welti, 67–80. Basel: Springer.

Gregg, M., and G. J. Seigworth, eds. 2010. *The Affect Theory Reader*. North Carolina: Duke University Press.

Gross, M. 2010. *Ignorance and Surprise: Science, Society, and Ecological Design*. Cambridge: MIT Press.

Hacking, I. 1983. *Representing and Intervening*. Vol. 279. Cambridge: Cambridge University Press.

Hacking, I. 1999. *The Social Construction of What?* Cambridge: Harvard University Press.

Haraway, D. 1988. "Situated Knowledges: The Science Question in Feminism and the Privilege of Partial Perspective." *Feminist Studies* 14 (3): 575–599. doi:10.2307/3178066.

Huutoniemi, K. 2016. "Interdisciplinarity as Academic Accountability: Prospects for Quality Control across Disciplinary Boundaries." *Social Epistemology* 30 (2): 163–185. doi:10.1080/02691728.2015.1015061.

Jarmusch, J. 2009. "Archival Talks: Jim Jarmusch, 'The Limits of Control'." *Film at Lincoln Center*. Accessed June 2019. https://www.youtube.com/watch?v=v62aGp0v64k

Kerr, H., and N. Nettlebeck, eds. 1998. *The Space Between: Australian Women Writing Fictocriticism*. Perth: University of Western Australia Press.

Latour, B. 1986. "The Powers of Association." In *Action, and Belief: A New Sociology of Knowledge?* edited by J. Law, 264–280. London: Routledge & Kegan Paul.

Latour, B. 2004. "Why Has Critique Run Out of Steam? From Matters of Fact to Matters of Concern." *Critical Inquiry* 30 (2): 225–248. doi:10.1086/421123.

Latour, B., and S. Woolgar. 1986. *Laboratory Life. The Construction of Scientific Facts*. Princeton: Princeton University Press.

Law, J. 2016. "Modes of Knowing: Resources from the Baroque." In *Modes of Knowing: Resources from the Baroque*, edited by J. Law and E. Ruppert, 17–59. Manchester: Mattering Press.

Law, J., and E. Ruppert. 2013. "The Social Life of Methods: Devices." *Journal of Cultural Economy* 6 (3): 229–240. doi:10.1080/17530350.2013.812042.

Lezaun, J., F. Muniesa, and S. Vikkelsø. 2013. "Provocative Containment and the Drift of Social-scientific Realism." *Journal of Cultural Economy* 6 (3): 278–293. doi:10.1080/17530350.2012.739972.

Lynch, M. 2017. "STS, Symmetry and Post-truth." *Social Studies of Science* 47 (4): 593–599. doi:10.1177/0306312717720308.

Maasen, S., and W. Peter, eds. 2005. *Democratization of Expertise?: Exploring Novel Forms of Scientific Advice in Political Decision-making*. Vol. 24. Dordrecht: Springer Science & Business Media.

Marres, N. 2018. "Why We Can't Have Our Facts Back." *Engaging Science, Technology, and Society* 4: 423–443. doi:10.17351/ests2018.188.

Mersch, D. 2015. "Rezeptionsästhetik/Produktionsästhetik/Ereignisästhetik." In *Künstlerische Forschung: Ein Handbuch*, edited by J. Badura, S. Dubach, A. Haarmann, D. Mersch, A. Rey, C. Schenker, and G. Toro, 49–57. Zürich: Diaphanes.

Oreskes, N., and E. Conway. 2010. "Defeating the Merchants of Doubt." *Nature* 465 (7299): 686–687. doi:10.1038/465686a.

Pickering, A. 2002. "Cybernetics and the Mangle: Ashby, Beer and Pask." *Social Studies of Science* 32 (3): 413–437. doi:10.1177/0306312702032003003.

Rhodes, C. 2015. "Writing Organization/romancing Fictocriticism." *Culture and Organization* 21 (4): 289–303. doi:10.1080/14759551.2014.882923.

Serres, M. 1981. *The Parasite*. Baltimore: John Hopkins University Press.

Stehr, N. 2001a. "A World Made of Knowledge." *Society* 39 (1): 89–92. doi:10.1007/BF02712625.

Stehr, N. 2001b. *The Fragility of Modern Societies: Knowledge and Risk in the Information Age*. London: Sage.

Thompson-Klein, J. 2004. "Prospects for Transdisciplinarity." *Futures* 36 (4): 515–526. doi:10.1016/j.futures.2003.10.007.

Weingart, P. 2001. *Die Stunde der Wahrheit? Zum Verhältnis der Wissenschaft zu Politik, Wirtschaft und Medien*. Weilerswist: Velbrück Wissenschaft.

Wilkie, A., M. Savransky, and M. Rosengarten. 2017. *Speculative Research: The Lure of Possible Futures*. Abingdon: Routledge.

Untrol: Post-Truth and the New Normal of Post-Normal Science*

Katharine N. Farrell

ABSTRACT

The idea that there exists a natural relationship between intellectual freedom, legitimate political authority and enjoyment of a dignified life was central to the European Enlightenment and to the radical social change it inspired. This normative claim was rooted in an epistemological proposition: truth is not revealed in private to a select few but discovered in public, through observation, dialogue and critique. Material transformations associated with that proposition have since literally changed the face of the earth. While the *materially* transformative potential of this proposition has been realized across the planet, its *social justice* implications have not. This leaves an underdetermined space in democratic discourse: fact claims are treated as apolitical, while the causes and consequences of the Anthropocene are uncertain and values regarding its importance polarised, rendering that status obsolete. In addition to contributing toward understanding human-environment relationships, fact now also serve to destabilize political discourse. Their instrumentalization exerts control in the absence of normative intention, 'untrol': truth claims matter not only because they call the moral subject to action but also because they can proffer political standing. Humbly embracing the epistemological complexity of the Anthropocene through Eagleton's posture of 'hope without optimism' is proposed as an antidote.

The Context and Conditions of 'Untrol'

Defining the present as a new geological epoch, the Anthropocene (Crutzen and Stoermer 2000; Waters et al. 2016; Zalasiewicz et al. 2011), can be understood as a normal science response (Kuhn [1962] 1970) to a post-normal science situation (Funtowicz and Ravetz 1991, 1992, 1993, 1997; Farrell 2011b): a normal-science response to extensive and robust evidence that the impacts of recent (in geological terms) anthropogenic interventions in dynamics of planetary change constitute a defining geological feature in earth's history; a post-normal science situation, in that the question of how to mediate, mitigate and ameliorate associated negative impacts, for humans and for life on this planet (the 'how-to' question of the Anthropocene), is an archetypical Wicked Problem (Allen et al. 2001; Rittel and Webber 1973, see also Farrell 2011b; Funtowicz and Ravetz 1993; Gibbons et al. 1994; Kates et al. 2001): persistent, complex and impossible to solve, due to its socially constructed character and its structural complexity, in technological (Rittel and Webber 1973), epistemological (Cilliers 2005; Farrell

*This was originally published as "Untrol: Post-Truth and the New Normal of Post- Normal Science" by Katharine N. Farrell in *Social Epistemology*, volume 34, issue 4 (2020), pp. 330–345.

2011a; Farrell et al. 2007; Marcuse [1964] 1991; Topping et al. 2015) and historical terms (Acosta García and Farrell 2018; Acosta García et al. 2017; Farrell 2008, 2011b; Ravetz 1971).

The phenomenon of post-truth (Fuller 2018), where facts are invented and employed for political as opposed to epistemological purposes, manifests clearly, although by no means exclusively, in post-normal science situations. While not identical, post-normal science and post-truth politics are intimately related. Both are symptoms of the near-sacred status of scientific facts in modern society. This makes the conceptual context of post-normal science situations, their historical origins (Farrell 2011b) and political implications (Farrell 2008, 2011a), a helpful point of reference for exploring post-truth. When considering the social character of these situations, we may distinguish between two types, differentiated by the way facts are subjected to validation. Following Farrell (2011b), who builds on Ravetz (1971), we can speak of first- and second-order situations. First-order situations are intentional collaborations, dedicated to verifying facts, undertaken *within* the academe, across disciplines, where the complexity of a problem indicates the need to employ a range of more or less ontologically and/or methodologically compatible technical expertise (Farrell 2011b; Farrell, Luzzati, and van den Hove 2013): e.g. predicting changes in the flow patterns of a river due to likely changes in exploitation. While this has political implications, the aim is a technical matter of concern (Latour 2004). Second-order situations may be intentional *or* circumstantially collaborative exercises dedicated to verifying facts, which extend beyond the academe, across disciplines and bridging scientific and political debate (Farrell 2011b): e.g. deciding rules of sustainable use for the same river.

Post-truth is taken here to be associated with second-order situations, where the complexity of matters of concern implies the need to employ both political (Barber 1984, 166) and scientific epistemologies (Popper [1959] 1968). Here, politics proper, what Mouffe (2005) calls The Political, becomes methodologically relevant to the work of fact validation (Farrell 2008, 2011b): because incommensurabilites in the ontologies and methodologies of those involved make non-political discourse impossible (Farrell [2005] 2009; Farrell, Luzzati, and van den Hove 2013; O'Neill 1993, 126); or because normative political positions, concerning the degree to which one or another fact is stable (e.g. anthropogenic climate change) are explicitly advanced; or both. It is not so much a choice, as an historical phenomenon (Fuller 2018), arising within a very particular context: the European Enlightenment (EE) project of developing rational, analytical modern science and the Industrial Revolution of the 19th century. This is a culturally entailed, historically embedded, social and ideological process, motivated by an ambivalent combination of aspirations to valorise both freedom and control (Horkheimer and Adorno [1944] 2006). Its treatment of facts as sacred is not a side issue but a structural feature of a cosmology intended to elevate rational analysis above theological caprice (Toulmin 1990; Camus [1951] 1953). As Camus points out, escaping the epistemological hold of theocratic caprice in the European Middle Ages was achieved by replacing one claim to definitive authority with another. However, faith in the veracity of facts identified using scientific method is still exclusionary (Stengers 1997, 110–118), and as Merchant (1981, 2006, 2008) reminds us, was never an innocent endeavour.

Today anthropogenic appropriation of river flows, nutrients cycles and territory has literally changed the face of the earth, while globalized communication infrastructure, which facilitates real-time, planet-wide distribution of information and misinformation, has changed how humans make sense, individually and collectively, of these changes. While the economies of India and China are increasingly major contributors toward the changes that characterize the Anthropocene, both its material and cultural impacts are euro-descendent (Malm and Hornborg 2014; Moore 2016).[1] This presents the euro-descendent elite, whether in Europe or elsewhere, with a choice: either (i) take responsibility for having fostered the present situation or (ii) endeavour to avoid that reality. The former implies seeking ways to address the irrevocably underdetermined 'how to' questions of the Anthropocene. This requires both humility and caution, because the one thing that is certain is that we cannot be absolutely certain about what is happening (Cilliers 2005). We might call this response, following Eagleton (2015a, 2015b), 'hope without optimism,' implying humility in the face of the real, a sense of responsibility for the past and moral duty toward the future. In distinguishing between hope and optimism, Eagleton (2015a) anchors hope in a critical attentiveness to the reality of circumstances, distinguishable from fantasy and lies. Optimism, in contrast, he associates with the emotive, not with anticipation but expectation, rendering it indistinguishable from charlatanism. Hope can be understood as a commitment to engage humbly with the real: 'an irreducible residue that refuses to give way, plucking its resilience from an openness to the possibility of unmitigated disaster' (Eagleton, 2015a, 114). 'Hope without optimism' is messy and uncertain; committed to keeping close to a reality that is also messy and uncertain.

The second response shifts attention away from euro-descendent cultural responsibility for the Anthropocene, exploiting the confusion and ambiguities of its wickedness. In place of the anticipated but impossible-to-create concrete propositions, alternative propositions, concerned with unrelated but concrete matters are put forward. That response is labelled here 'untrol,' to reflect its paradoxical character, where fact claims serve to disrupt collective sense-making efforts. It reflects a post-moral politics, where the instrumental value of scientific facts, which helped to make modern industrialization possible, is fetishized, and fact claims become instruments of coercion. Responding to the 'how to' questions of the Anthropocene with 'hope without optimism' implies collaborative research that extends across conventional academic disciplines, across coloniser/colonised divides and bridges political, cultural and scientific discourses. Responding with 'untrol' implies simplification, polarisation and walling apart political, cultural and scientific discourses. 'Hope without optimism' is characterised by epistemologically complex discourse (Farrell [2005] 2009), where what is understood to be true and real is based on reflected upon experience (Dewey [1929] 1958) and negotiated through discourse and dialogue (Fuller 2003, 2018; Habermas 1984; Kant [1787] 1990, [1784] 1991). 'Untrol,' is characterised by an 'overwhelming concreteness ... an irreconcilably anti-critical and anti-dialectical language' (Marcuse [1964] 1991, 98–100), where the identification of objects with their function erases their relationship to the concepts and subjects that define them, eliminating the point of reference for moral accountability (98–99).

Historical Relationship between 'Untrol' and the Phenomenon of Wicked Problems

The idea that there is a natural relationship between truth, authority, freedom and enjoyment of a dignified human life inspired the intellectuals of the European Enlightenment (EE). Their early discourses were comprised of questions concerning both epistemology and ethics: with determining the capacity of reasonable beings to know, through observation, reflection and free and open discourse, what is true (Kant [1784] 1991). Their normative call for the freedom to think independently, advanced with great fervour by the French *Jacobin* movement, was central to the EE epistemology out of which the modern Scientific Method arose (Fuller 2003; Popper [1959] 1968; Ravetz 1971). That is to say, there is an historical relationship between the scientific achievements of the EE and the idea that human rationality is the appropriate referent for ensuring the good life (Camus [1951] 1953; Toulmin 1990). The consequences of this epistemological shift are well documented and were a matter of concern for a number of philosophers in the 19th and 20th Centuries, among them (Goethe [1808–1832] 1938, Nietzsche 1882, Dewey [1929] 1958 Popper [1959] 1968, Horkheimer and Adorno [1944] 2006), Marcuse (1969, 1972[1955], 1991 [1964]) and Arendt (1958, 1963, 1970).

The normative proposition of the EE, exemplified by the French motto *liberté, égalité, fratenité*, reflected an effort to escape the intellectually suffocating grip of religious dogma and associated authoritarianism in the European Dark Ages. The ensuing Industrial Revolution was situated within this European cultural history (Camus [1951] 1953; Horkheimer and Adorno [1944] 2006; Toulmin 1990) from which the EE cosmology that supports the emergence of 'untrol' cannot be easily extricated. However, the egalitarian ideals espoused by the *Jacobins* and their contemporaries challenged the authority of an economic system on the verge of achieving global influence. That accident of history precipitated a co-determination of discourses and identities between colonizer and colonized (Mignolo 2000), where the cosmologies, customs and sense-making practices of the colonized were necessarily relegated to the status of 'an *other* rationality' (Stengers 1997, 110), because euro-descendent modern science claimed to be engaged in determining universal truths.

While there were surely a range of contributing factors, it is clear that, during this transitional period, an incomplete and perverted translation of the highest ideals of EE was projected onto the world stage.[2] What was a radical proposition in Europe – that all (hu)men (beings) have a basic right to lead what they themselves deem to be a dignified life – was combined with a culturally attenuated discourse concerning what constituted a dignified life. Egalitarianism at home, became, in the colonial sphere, authoritarian repression of non-aligned cultures (De Beukelaer and Fredriksson 2018; Murdock 2018). This intimate relationship between a myopic vision of egalitarianism, taken to mean access to sameness (Woltersdorff 2003), and a morality of free and open critical debate in science, is part of the history of the political economy of modern science (Bahro 1977; Farrell 2008, 2011a; Marcuse 1978, [1964] 1991; Mirowski 2011; Saltelli 2018). The justification for oppression of and disdain for the cultures of colonized and enslaved peoples and their epistemologies was that they were technologically inferior. However, the historical euro-descendent step of separating truth from the divine, which created the epistemological space required to develop the Scientific Method and to invent associated industrial technologies (Nietzsche 1882), also brought with it a shift in European social structures, which, unable to derive their authority from divine right, instead derived it, from then onward, based on a democratic mandate and epistemological veracity (Camus [1951] 1953; Horkheimer and Adorno [1944] 2006): moral questions became, ostensibly, a political matter, and 'how-to' questions were to be addressed by scientists and technical functionaries of the state (Arendt 1963, 1970; Bahro 1977; Weber [1904] 2015).

The EE social contract between science and society, as Ravetz (1971, 1990) describes it, is the source of scientists' authority to speak in modern democratic discourse. In this respect, it is also the origin of the political power employed by the practitioner of 'untrol.' Arendt describes the EE roots of this authority as follows: 'Only with the rise of Puritan morality, coinciding with the rise of organized science, whose progress had to be assured on the firm ground of the absolute veracity and reliability of every scientist, were lies considered serious offenses.' (Arendt 1967, 50). This reflects the euro-descendent cosmological shift documented in real-time by Nietzsche and in retrospect by Camus and Toulmin, from belief in divine authority, to belief in the authority of what Arendt refers to as 'organized science.' She addresses the political implications of these new, post-aristocratic social structures using the concept of a technocratic rule of Nobody.

> In a fully developed bureaucracy there is nobody left with whom one could argue, to whom one could present grievances, on whom the pressures of power could be exerted. Bureaucracy is the form of government in which everybody is deprived of political freedom, of the power to act; for the rule by Nobody is not no-rule, and where all are equally powerless we have a tyranny without a tyrant. (Arendt 1969, §V)

Under the rule of Nobody, the moral position of egalitarianism espoused by the *Jacobins* has no purchase, while the materially transformative capacity of the EE epistemology continues at full momentum. This is further complicated by what Ravetz (1971, 1990, 2013) refers to as the industrialisation of scientific knowledge production, where scientists are now, in spite of retaining their political privilege to speak truth to power, ostensibly paid wage-labourers (Bahro 1977; Farrell 2017; Marcuse 1978). The inherent indeterminacy of The Real (Wakefield 1990) undermines scientists' authority to speak as representative of the truth and their socio-economic status transforms them into instruments of political economy. What remains unchanged is that facts are produced which afford standing in political discourse. That they are inscribed, as Latour and Woolgar (1979) put it, is taken here as given: not in the sense that they are invented from thin air but rather that they are crafted, through careful and rigorous attention to the real world. However, at the frontiers of modern science, the distinction between fact and opinion is blurry. The theory dependence of instruments (Chalmers 2003) leads to a situation where the systems used to structure what is knowable (epistemologies) literally delimit what is understood to be (ontology): models can and do produce reality (Wakefield 1990). Dealing with this conflation of ontology with epistemology implies comprehending the drivers of the Anthropocene *from within* (Bahro 1977; Blühdorn 2000; Horkheimer and Adorno [1944] 2006; Stengers 1997); producing facts becomes inescapably political.

Following Camus [1951] 1953, we can distinguish between two styles of response to this situation: Nietzschean nihilism, where the self-referential dynamic associated with the anthropogenic character of the apparatus used to make sense of the causes and consequences of the Anthropocene is viewed as a condemnation to semiotic irrelevance; and a Deweyan pragmatism, characterized by humility and precaution, in which epistemology is presumed to have always been constituted in this manner and the knowable limited to the now. The former, eschewing the epistemological complexity of Wicked Problems, leaves democratic discourse intrinsically underdetermined, creating space for the opportunistic instrumentalization of fact claims (Blühdorn 2000); the latter, implies substantive reform in the rules of democratic political engagement, replacing the Athenian presumption of a universal subject with recognition of the many voiced, agonistic discourse that characterizes The Political of the Anthropocene (Farrell [2005] 2009; Latour 2004; Machin 2019; Mouffe 2005; Stengers 1997).

How the Style of 'Untrol' Manifests

An opening move of the 45th President of the United States (hereafter, sitting US President), the dissemination of what Kellyanne Conway, an adviser to him, famously referred to as 'alternative facts,' illustrates how 'untrol' is practiced. Then White House Press Secretary Sean Spicer's claim, that the number of people attending the January 2017 inauguration was historically high, was widely rebuked (Hunt 2017). Conway defended it based on the claim that the White House had 'alternative facts' (Blake 2017). While perhaps a matter of personal pride for the sitting US President, or of positional importance for the Republican party, the question of how many people attended the inauguration is, ultimately, trivial. That it is trivial is, however, anything but. Rather, it highlights the instrumental, as opposed to dialogical, character of 'alternative facts.' It is not their content, but their rhetorical status that counts: as claims to speak from the moral high ground of facticity, they enjoy an authority imputed by the euro-descendent cosmology that underpins sense making in globalized late-industrial society. This status authorises their speaker to enter a political discourse characterised by what Marcuse ([1964] 1991, 98) has called an 'overwhelming concreteness', where 'real' replaces 'good' as the moral point of reference (Acosta García and Farrell 2018; Ariza-Montobbio and Farrell 2012; Farrell 2008, 2011b):

> The distinction between desirable and undesirable is replaced by a distinction between rational and irrational. Rational is associated with the production of facts, the success of science and the conquest of nature. It is implicitly treated as good and desirable. Irrational, defined only through negation, is the absence of rationality and is associated with subjugation to the forces of nature. It is implicitly treated as "other" from good and not desirable. (Farrell 2008, 77)

Where 'alternative facts' are employed, we are *in* a manifestation of Latour's (2004) Parliament of Things: standing is awarded based on nominations by authorised members of the existing collective: 'facts.' The post-truth posture adopted by such nominees, exploits a form of political power already at the disposal of the purveyors of 'conventional facts.'[3] 'Untrol' gains purchase, where realism cannot because, in post-normal science situations, the truth is that we don't know what the truth is (Allen et al. 2001; Farrell [2005] 2009; Pellizzoni 2003; Machin 2019; Ravetz 1971, 1990). In this context only facts matter. Moral questions, for example, those associated with accusations of criminal and immoral activities on the part of the sitting US President, are peripheral; attention is focused on the concrete realities of who said what and whether or not it was true. This style of amoral politics is possible where there is an epistemological climate of overwhelming concreteness, where impertinent but concrete matters of concern are prioritised over pertinent but abstract ones (Ariza-Montobbio and Farrell 2012). As Marcuse ([1964] 1991) points out and Eagleton (2015a, 2015b) reiterates, this can be understood as a contemporary vestige of the violent separation of the moral from the real, of facts from feelings, that characterized the EE. In rendering facts amoral it was possible to control forces, such as electricity, that had been mysteries of the heavens. Merchant (2008, 734) describes the project and its material price as follows: 'Bacon's ultimate objective was to recover the "dominion over creation" lost in the Fall from Eden in order to benefit humanity in material terms. That dominion, however, was achieved by the constraint of nature through technology, a process that exacted heavy costs from nature itself.'

Anthropogenic climate change and mega-contamination are judged by reputable experts to be on the verge of spinning out of control (IPCC 2018; Steffen et al. 2018) and there are now, across the planet, numerous anthropogenic biological dead- and no-go zones, including Chernobyl and Fukushima, over 1,000 superfund sites in the US, numerous maritime dead zones (Diaz and Rosenberg 2008; Schmidtko, Stramma, and Visbeck 2017) and several ocean vortex plastic islands (Cózar et al. 2017), to mention but a few examples. Positions that downplay or ignore this reality cannot be defended through reference to the truth, in either the ratio-logical or the experiential sense. However, as their standing in political discourse concerning 'how to' questions of the Anthropocene depends upon their facticity, they are dressed up for the part. Since veracity can only be attributed to their status as opinions – i.e. it is true that this is my opinion – the existence of an opinion is proffered as an opinion regarding the facts – i.e. an 'alternative fact.' Within democratic discourse structured through the EE social contract between a moral body politic and an amoral science, such cross-dressing produces chaos. Claims that are neither explicit moral propositions nor plausible representations of scientifically verified positions cannot be held to account in a discursive context equipped to deal only with one or the other.

In addition, the Anthropocene is characterized by changes in how the relationship between humans and their environments is perceived by humans. Today, big data and deep machine learning help shape how human-environment relationships are perceived and experienced, generating further epistemological instability. The private company Alphabet Inc., which reported interim 6-month revenues of nearly $64 billion dollars for the second quarter of 2018 (Alphabet Inc 2018), provides its Google search engine service for free, selling to its paying customers access to its non-paying user and information about them, while simultaneously co-constructing the representations of reality that these users reference, through proprietary algorithms (Couldry, Fotopoulou, and Dickens 2016) that, presumably, serve to structure content in ways that serve the interests of their paying customers. These instrumentalized and interest laden Google versions of reality are what millions of humans today use to navigate their way through the environments within which they live: to discern the real. Boundaries between artificial intelligence and human cognition are also materially blurred. The state of Saudi Arabia has given citizenship to a humanoid robot, Sophia (Arab News 2018), and neurocognitive machine interfacing, in the form of neuroprostheses, is now a basic feature of spinal cord injury treatment (Chapin et al. 1999; Creasey et al. 2001; Glannon 2016; Solinsky and Specker Sullivan 2018; Warwick et al. 2003). The ubiquity of smart phones, which both collect and provide Google´s products, having reached market saturation at an estimated 1.5 billion in sales in 2017 (Mayeda 2018), implies a generalised shift in the structural basis of collective human cognition, toward algorithm-based filtering, tailored to micro-communications. The Twitter-based communications of the sitting US President fit well with the profile; this new structural context favours the practice of 'untrol,' where simple fact claims, regardless of their veracity, are communicated in short snippets, while the complex underdetermined substantive signifieds of the Anthropocene are hard to convert into soundbites, and so less prevalent in this epistemological space.

Reaching back to the philosophical origins of the post-normal science discourse (Farrell 2011b) sheds some light on why it has proven so difficult to establish overarching standards for quality assurance in science concerning the Anthropocene and so easy to advance political discourses based on 'alternative facts.' In his 1971 text *Scientific Knowledge and its Social Problems*, Ravetz describes what he later refers to as extended peer communities (Funtowicz and Ravetz 1991), which have social standing to decide whether a proposition is considered a fact. He argues that the industrialization of scientific knowledge production, once conceived of as a noble project of the European elite, has changed the relationship between patrons and the technical content of scientific work. Under industrialized conditions, propositions become facts based not only on method but also on the preferences of patrons, who decide which problems are studied and which methods employed. Where collaborations are between scientists of different disciplines, while the work from one field may be of interest in another, individuals lack the ability to judge its technical quality. In these situations, described above, respectively, as second- and first-order extended peer-review contexts, judgements, about what constitute the facts of the matter, are political (Farrell 2008, [2005] 2009, 2011a) and the combination of disciplinary differentiation, transdisciplinarity and the sheer volume of scientific output render customs of 19th Century peer-review obsolete (Ravetz 2013).

Applying this frame, the operating context of 'untrol' can be broken down into nine basic situations (Figure 1), each implying a degree and type of exposure to the practice. The matrix to the left in Figure 1 represents implications for scientists; the matrix to the right, implications for the body politic. Box 7, in the bottom left corner of both matrices, represents conditions under which normal science can be practiced: decision stakes and system uncertainty are both low. Traditional standards of peer-review and generalizability of results apply. Boxes 4 and 8 represent what Funtowicz and Ravetz (1991) refer to as consultancy situations, where a matter of concern implies raised decision stakes or raised system uncertainties: a bridge might collapse, causing harm to people and property; disease might spread. In these situations, there is scope for contesting facts but they still enjoy a good degree of stability. Boxes 1, 2, 3, 5, 6 and 9 reflect post-normal science situations, where facts become unstable and 'untrol' may be practiced. In boxes 1 and 2 science is not contested but selecting the matter of concern is. In debates concerning genetically modified foods, one may focus on the human health implications of consumption or the ecological impacts of cultivation. Neither aspect is scientifically underdetermined but political contestation regarding which question matters generates epistemological instability: is the technology really safe? In boxes 6 and 9, contestation centers on choosing between a range of competing, uncertain scientific propositions, to address a non-contested political aim. For example, in spite of political consensus that carcinogens should not be introduced into drinking water, the EE methodological proposition that carcinogenicity should be evidenced through statistically significant indicators of causal relations obviates considering errors associated with the omission of stochastic and relational variables relevant for characterizing change dynamics in living systems (Prigogine and Stengers 1984). In boxes 5 and 3, there are weak or no stable scientific referents available and only weak or no points of political consensus: in box 5, careful attention to detail and patience may help committed social actors identify facts relevant for sense making; in box 3, all bets are off.

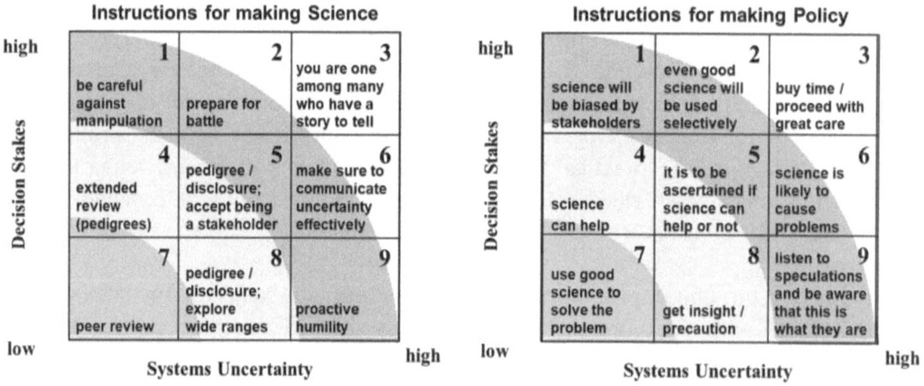

Figure 1. The anatomy of destabilized fact claims under post-normal science conditions.

Source: (Farrell et al. 2007). Adapted from schematics presented in (Funtowicz and Ravetz 1991, 1992, 1993), based on reference to (Ravetz 1971).

When 'how-to' questions of the Anthropocene are addressed collectively, the combination of unstable facts and the moral status of 'that which is true' being 'that which is right' (Farrell 2011b) creates an epistemologically unstable situation. This, combined with the standing afforded facts under an EE epistemology, makes possible the practice of 'untrol' because, as discussed above, the *Jacobin* claim to a right to rule was not only normative but also epistemological (Camus [1951] 1953; Gould 2003). Without wishing to suggest that Kuhnian normal science is irrelevant for addressing the challenges of the Anthropocene, debate concerning 'how-to' questions is irretrievably post-normal science: facts are underdetermined and values are polarised, puzzle solving does not make for good science (Allen et al. 2001) and peer-review is *de facto* extended (Farrell 2011b). Professionally, culturally and cosmologically diverse communities, including scientists, collaborate in the work of determining what count as the facts of the matter (Farrell 2011b; Latour 2004; Ravetz 1971).

Ironically, in its openness, the EE Scientific Method creates room for 'untrol.' As Cilliers (2005) points out, while there is nothing more logically water tight than a tautological statement, complex problems do not lend themselves to tidy solutions. This obliges the attentive student of the Anthropocene to tolerate a degree of ontological and epistemological indeterminacy. The question as to how one works with this indeterminacy, whether it is exploited for instrumental purposes, or regarded and respected, distinguishes practices of 'untrol,' characterized by the former, from 'hope without optimism,' characterised by the latter. The force of irrefutable tautological statements in the post-truth discourse is illustrated by then Judge, now Justice Brett Kavanaugh's inane response to the question of whether or not he drank to excess and, might, as a result, not be able to remember having assaulted his accuser, Dr. Christine Blasey-Ford. His emphatic and repeated response, 'I like beer,' was impossible to refute. It was also entirely irrelevant. While the matter of concern was not a 'how-to' of the Anthropocene, the political climate of Kavanaugh's confirmation sits squarely in box 3 of Figure 1: framed by conflict between outgoing and incoming administrations, in which the decision of the latter to pull out of the Paris Climate Agreement has been a central element.

Another example of the 'untrol' posture, can be seen in the autumn 2018 response of the then sitting US President to allegations that Jamal Khashoggi was murdered in the Saudi Arabian Embassy, in Istanbul. Khashoggi, a Saudi citizen, permanent US resident and journalist, was, in the preceding months, increasingly critical of the Saudi regime (Ignatius 2018) and there have been indications, including from within the US Intelligence Service, that the Saudi government was directly involved in his death (Zengerle 2018). The sitting US President, however, cautioned that judgements regarding this matter of concern should be measured against considerations that the Saudi government was negotiating a multi-billion dollar arms purchase from US suppliers (Borger and Chulov 2018; The Guardian 2018; White House 2018). Here the questions with moral significance – whether or not Mr. Khashoggi was indeed murdered; if so by whom; and what does this imply for human rights – were displaced, rhetorically, by a fact claim: the US has a pending arms trade agreement with the Saudis.

While the question as to whether or not Mr. Khashoggi was murdered for political reasons at the behest of an allied government is an obvious matter of political concern, the sitting US President's response was that it mattered only insofar as it did not disrupt the status quo. Bereft of moral sentiment, this response achieves moral status through an appeal to the real, to the way things are. While there was surely an element of Real Politik, the content and character of the sitting US President´s statements illustrate the instrumental, as opposed to dialogical role that fact claims played. In lieu of owning the Real Politik hypocrisy of tolerating the assassination, the White House tried to push the narrative into box 9, by invoking the underdetermined character of knowledge concerning what really happened, then shifted the focus to a concrete but logically unrelated fact '... we may never know all of the facts surrounding the murder of Mr. Jamal Khashoggi. In any case, our relationship is with the Kingdom of Saudi Arabia.' (White House 2018).

Responding to critiques of dubious fact claims with robust but off topic facts can be understood as a form of 'untrol.' It is a core feature of the public commentary styles of both Conway and Sarah Huckabee Sanders (Mr. Spicer's now retired replacement) and can be observed in the ample video documentation of their many public appearances. In place of the customary moral posture of defending allies for strategic reasons, the moral posture of the White House, with regard to the assassination of Khashoggi, was opportunist fatalism: what 'is' is prioritized over what 'ought to be' (Ariza-Montobbio and Farrell 2012; Marcuse [1964] 1991). This 'is'-oriented morality is a basic feature of 'untrol:' lacking the aspiration to exercise control, it is banal and insidious, as Arendt (1963) argued with regard to Eichmann, because it is unthinking. Abdicating responsibility to reflect upon what motivates one to act, the exercise of 'untrol,' through opportunistic exploitation of factual ambiguities, seeks discursive position, as an end itself, while the associated destabilizing of democratic discourse, directed here not by intent but opportunism, serves to reinforce the very epistemological instability that it exploits.

In late-industrial political discourses defined by overwhelming concreteness, being in command of the facts affords one authority (Ariza-Montobbio and Farrell 2012; Farrell 2008, 2011a; Marcuse [1964] 1991). The comportment of Conway in an interview with CNN commentator Chris Cuomo on 24 August 2018, also readily available online, illustrates the style: accusatory, reactive and dismissive, she persistently and repeatedly responds to critique with unrelated fact claims. Sanders, whose briefings are also freely available on the internet, had a similar, if less ostentatious style: unwelcome questions are answered with unrelated fact claims. Her 3 October 2018 Press Conference, where she fielded questions concerning Blasey-Ford's testimony before the US Senate and questions concerning the sitting US President's comments about it, is literally filled with off topic assertions of facticity: there is no moral issue at hand; the facts do not support Blasey-Ford's accusation; the sitting US President was merely stating the facts when questioning the veracity of Blasey-Ford's accusations. The complementary fact, that a victim's memory of rape or attempted rape is usually distorted and often does not meet the standard for evidence in criminal prosecutions (Hohl and Conway 2016), is selectively disregarded. Here we are situated in box 1 of Figure 1, where a political choice is made regarding which set of stable facts to invoke in a discourse. That there were gaps in Blasey-Ford's recollections, is a stable fact. Its selection for emphasis, over other stable facts concerning the complexity of trauma memory was political, serving to direct discourse away from a moral matter of concern: is Kavanaugh fit to serve as a Justice?

Ethical Implications of "Untrol:" Anthropocene Science and "Hope without Optimism"

I take part of the title of this last section from Terry Eagleton's (2015a) book *Hope without Optimism*, which was recommended to me by a Turkish scholar, living outside the country, as we discussed the implications of constitutional reforms that had just been assigned by Recep Tayyip Erdoğan's near monarchical authority, in 2017. Incarceration of journalists had become widespread in Turkey and international travel of academics was being monitored. Eagleton's general proposition is that the expectation of things turning out for the best actually mitigates against constructive engagement with how things are at present. It is in that spirit that I close this text, with an appeal to 'hope without optimism,' gesturing toward a style of post-truth politics that can counter-balance 'untrol.'

Eagleton's phrase calls attention to what Fuller has called an 'intellectual responsibility' (Fuller 2003, 17), left untethered by Kuhn [1962] 1970 in the argumentative acrobatics of the 1970 postscript to his text *The Structure of Scientific Revolutions*. There Kuhn attempts to distance himself from the post-modernist implications of his position regarding the socially mediated character of scientific progress. Rather than attempting to argue that scientific facts are either immutably real or pure constructs, a posture of 'hope without optimism' is empirical but provisional, conscious of its epistemological limitations and open for things to turn out differently. In this respect it has a strong resonance with the spirit of '[t]he open society … whose members, like the citizens of classical Athens, treat openness to criticism and change as a personal ethic and a civic duty.' (Fuller 2003, 26). However, a democratic forum equipped to facilitate debate on the 'how-to' questions of the Anthropocene would look quite different from the euro-descendent Athenian ideal of one man, one vote, within which 'untrol' is able to make itself at home. It would provide room for the epistemological complexity of a socially, economically, culturally and professionally diverse body politic, where diverse interventions by subjects with differentiated experiences of, knowledge concerning and rights with respect to the causes and impacts of the Anthropocene do not serve consensus but comprehension. One way to achieve this might be a sort of corporatist federalism (Farrell 2004, [2005] 2009), with standing linked to social roles and relationships to the causes and impacts of the Anthropocene. Machin (2019, 354) recommends '[a]n agonistic approach to [this] incommensurability ... [which] would consist of respect for the diversity it brings and celebration of its fecund potential.' Although precisely how such changes might be configured is beyond the scope of the current text, Eagleton's concept of 'hope without optimism' offers a guide to conceptualizing the style: humble, focused on the present, with the aim of developing realistic aspirations for the future, in place of unrealistic expectations.

Staying with the example of the Anthropocene context in the US at present, we can see examples of this alternative response in the discursive styles of Blasey-Ford and the late Senator John McCain. McCain, who was known for being candid, made no secret of his disdain for the sitting US President's disposition to conflate opinion with fact. His political leadership, as an opponent of 'untrol' was clear at his funeral, to which the sitting US President was not invited and at which both his ally George W. Bush and his opponent Barack Obama presented eulogies. That McCain was enthusiastic about critique is exemplified in his preparation, upon returning from Vietnam, where he was a prisoner of war, of a set of critical comments on the US Military's Code of Conduct for Prisoners of War. Reviewing 'the Code of Conduct, article by article, ... [the report] assesses its value and viability as they related to the Vietnam experience' (McCain 1974, ii). His first recommendation, supported by a detailed review of the experiences of prisoners of war in the theatre, is humble: that the Code of Conduct, Article III, concerning escape, should be changed from 'I will make every effort to escape and to aid others to escape...', to read, 'I will make every reasonable effort to escape...' (13). The argument is not advanced from a posture of bravado but of humility and openness. Blasey-Ford's testimony, ridiculed in public by the sitting US President for its failure to proffer facts, provided clarity. In contrast to Kavanaugh's emotional outbursts and 'untrol' redirects – 'I like beer' – she adopted a firm but retiring posture, demonstrating humility in the face of uncertainties. The style of McCain and Blasey-Ford, is one of epistemological solidarity (Farrell and Barkin 2016): a compassionate regard for the limits of one's own knowing and the validity of other ways of making sense of reality.

Hope, Eagleton (2015a, 3) argues, is like love, 'It must be able to pick out the features of a situation that render it credible. Otherwise it is just a gut feeling ... Hope must be fallible, as temperamental cheerfulness is not.' In Anthropocene science and politics, it is an act of solidarity with those who would know what is real: a willingness to allow one's own truth to be affected by that of another (Brennan 2004). This implies attention to an ethical duty to engage with the forgotten social justice aspect of the EE epistemology, by treating all interlocutors, regardless of their cosmological disposition, as potential members of the sense-making collective. In this respect, it is also reminiscent of Dewey's [1929] 1958 empirical philosophy, concerned with formalising attention to the place of the observer in the act of observation. It invites a responsible posture with regard to the relationship between the EE cosmology and the momentum of the Anthropocene, where the inevitably politicized interpretive role of the observer, generating and employing scientific knowledge about the causes and consequences of the Anthropocene, is formally considered.

The exercise of 'untrol,' reflects an instrumentalist disregard for the dignity of truth and implies an ethical duty on the part of 21st Century scholars, to engage with the forgotten social justice aspect of the EE epistemology: that discovered truth is an equalizing force. The posture of 'hope without optimism' is needed today because, under the post-truth political conditions amenable to 'untrol,' speaking truth to power is not useless; it is impossible. With regard to the 'how-to' questions of the Anthropocene, there is no plausibly apolitical truth to be spoken. The idea that truth matters serves the practitioner of 'untrol' not because it reveals the composition of ethical dilemmas and moral choices, but because it can be used, instrumentally. It matters to the practitioner of 'hope without optimism' because the combined epistemological/ethical roots of the EE are worth defending.

The question as to how the transformative potential of EE derived technological transformations can be brought under control has been posed for decades (Carson 1963), with precious little to show for the effort. The advent of post-truth politics, explored here through reference to the concept of 'untrol,' marks an inward turn in what had already become a perverse process of self-referential recursive co-evolution (Farrell 2007) of expectations and beliefs (Bahro 1977), where the anthropogenically attenuated world of late-industrial life *is* the world to which human societies are adapting (Bahro 1977; Farrell 2017; Marcuse 1969, [1955] 1972, [1964] 1991). Continuing exercise of 'untrol' can be expected to lead political discourse concerning the Anthropocene away from the uncomfortable realities implied by impending cascades of ecological and geological system collapses (Steffen et al. 2018). Exercising 'hope without optimism' could, in contrast, lead to methodological innovations that improve the quality of understanding and actions. Bearing in mind that, as Lander puts it, '…the naturalization of liberal society as the most advanced and normal form of human existence is not a recent construction that can be attributed to neoliberal thinking, nor to the current geopolitical juncture, but instead has a long history in Western social thinking of the last centuries' (Lander 2000, 12), it stands to reason that the changes required to make space, in a world conducive to 'untrol,' for 'hope without optimism,' would not be superficial but structural.[4] 'Untrol' reflects a step forward, in EE history, from the scientific fact serving as instrument of a euro-descendent cosmology, to the instrumentalization of facts as mechanism of amoral politics. 'Hope without optimism' reflects a refusal to ignore the complexity of the Anthropocene.

In the Anthropocene, humanity is faced with a form of epistemological challenge that is not resolved within the euro-descendent cosmology that structures contemporary international trade, politics and academic enquiry. That challenge was anticipated by Nietzsche and discussed at length by Horkheimer and Adorno ([1944] 2006), Marcuse (1969; 1972[1955], 1978; 1991 [1964]) and Bahro (1977): the intellectual elite engaged in the production of organized scientific knowledge are the authors of the EE cosmology. As such, they are responsible, whether implicitly or explicitly, for valorising the current destructive dynamics of the Anthropocene. This implies a moral duty to make virtuous choices concerning the structure and content of the planet's hegemonic cosmology. Haraway's (1988, 2016) response, what she calls embodied objectivity, brings the discussion to a place that is compatible with Dewey's ([1929] 1958) epistemology and provides a basis for opening up the collective responsible for constructing this new EE cosmology, to a wider community. Both Dewey's empirical philosophy and Haraway's embodied objectivity resist the temptation to slip into a posture on one or the other side of what Cilliers (2005) calls the family fight between radical realism and radical relativism. Instead, they seek to formalise engagement with the inevitability of neuro-cognitive modulation in the act of observing (Dennett 1992).[5]

> So, I think my problem, and "our" problem, is how to have simultaneously an account of radical historical contingency for all knowledge claims and knowing subjects, a critical practice for recognizing our own "semiotic technologies" for making meanings, and a no-nonsense commitment to faithful accounts of a "real" world, one that can be partially shared and that is friendly to earthwide projects of finite freedom, adequate material abundance, modest meaning in suffering, and limited happiness. Harding calls this necessary multiple desire a need for a successor science project and a postmodern insistence on irreducible difference and radical multiplicity of local knowledges. (Haraway 1988, 579)

'The moral is simple: only partial perspective promises objective vision' (Haraway 1988, 583). While dominant euro-descendent political customs are still a long way from creating the discursive spaces required to afford standing to such partial perspectives, there are spaces opening up to the practice of 'hope without optimism,' which might eventually invite them in. One example is the introduction, in several recent state constitutions, of rights for nature; another is the openness to epistemological complexity reflected in the methodology adopted by the Intergovernmental Science-Policy Platform on Biodiversity and Ecosystem Services (IPBES) (Díaz et al., 2015; IPBES 2019). Ironically, where 'untrol' reflects the exercise of power in the absence of underlying intention to control outcomes, destabilizing collective sense making, 'hope without optimism,' in abandoning the expectation that outcomes can be controlled, represents a commitment to work with the complex reality of an epistemologically underdetermined Anthropocene, making possible informed choices that could ultimately enhance human self-determination.

Notes

1. In this text, the geological and social pertinence of the Anthropocene is taken as given. While critiques of the use of the term bring forward further issues related to intellectual colonialism, it is beyond the scope of this text to address them here in detail.
2. While it has been argued by Horkheimer and Adorno [1944] 2006 from a political philosophy perspective, and Merchant (1981), from a methodological one, that failure to realize the liberating potential of EE ideology is built into the logic of replacing faith in divine enlightenment with faith in reason, engaging with the subtleties of those arguments would bring us away from the present topic. While it is certainly important, eventually, to determine the cause of a fire that has brought down a building, at the moment of the fire itself, what is of utmost importance is to identify viable routes that one can use to escape.
3. This manifestation of Latour's Parliament of Things is a complicated and busy space, occupied, not only by our current companions but also by a broad array of non-human moral objects – from polar bears to diverted rivers – by a spectrum of feminist and queer subjectivities, and by a host of actors and matters of concern relating to the rights and dignity of colonised peoples across the planet, whose cosmologies and contexts do not fit the euro-descendent idea of progress. While these actors and matters of concern are assumed to play a part in the workings of both 'untrol' and 'hope without optimism,' it would be foolhardy to try to also unpack here their accompanying.
4. Original Spanish text of the quote from (Lander 2000, 12): ' … la naturalización de la sociedad liberal como la forma más avanzada y normal de existencia humana no es una construcción reciente que pueda atribuirse al pensamiento neoliberal, ni a la actual coyuntura geopolítica, sino que por el contrario tiene una larga historia en el pensamiento social occidental de los últimos siglos.'
5. Haraway, like Vandana Shiva (Shiva 1989) and Carolyn Merchant (1981), explicitly links EE culture gender discrimination with her epistemological position. Although that matter is not of immediate pertinence to the position being developed here, it should be noted.

Disclosure statement

No potential conflict of interest was reported by the author.

Funding

The research for this text was conducted under the auspices of the author's position as a staff member with the Department of Biology, Faculty of Natural Science and Mathematics, Universidad del Rosario and did not include reliance on any third party financial support.

ORCID

Katharine N. Farrell http://orcid.org/0000-0002-2096-3344

References

Acosta García, N., and K. N. Farrell. 2018. "Crafting Electricity through Social Protest: Afro-descendant and Indigenous Embera Communities Protesting for Hydroelectric Infrastructure in Utría National Park, Colombia." *Environment and Planning D: Society and Space* 37: 0263775818810230.

Acosta García, N., K. N. Farrell, H. Heikkinen, and S. Sarkki. 2017. "A Teleological Approach to the Wicked Problem of Managing Utría National Park." *Environmental Values* 26: 583–605. doi:10.3197/096327117X15002190708128.

Allen, T. F. H., J. A. Tainter, J. C. Pires, and T. W. Hoekstra. 2001. "Dragnet Ecology—"Just the Facts, Ma'am": The Privilege of Science in a Postmodern WorldScience of Intrinsic Quality Needs Narratives with Explicit Values—Not Just Facts—Particularly as It Faces Multiple-level Complexity in Advising on Environmental Policy, Such as Planning for Energy Futures." *BioScience* 51: 475–485.

Alphabet Inc. 2018. "Quarterly Report Pursuant to Section 13 or 15(d) of the Securities Exchange ACT of 1934, for the Quarterly Period Ended June 30, 2018." *Alphabet Inc.* Accessed October 2018. https://abc.xyz/investor/static/pdf/20180723_alphabet_10Q.pdf?cache=2e7508b

Arab News. 2018. "Saudi Arabia Becomes First Country to Grant Citizenship to a Robot." *Arab News*. Accessed October 2018. http://www.arabnews.com/node/1183166/saudi-arabia

Arendt, H. 1958. *The Origins of Totalitarianism*. Cleveland: Meridian.

Arendt, H. 1963. *Eichmann in Jerusalem: A Report on the Banality of Evil*. New York: Viking Press.

Arendt, H., 1967. "Truth and Politics." *The New Yorker*, p. 49.

Arendt, H. 1969. "A Special Supplement: Reflections on Violence." *New York Review of Books*, February 27.

Arendt, H. 1970. *On Violence*. London: Harcourt Brace & Company.

Ariza-Montobbio, P., and K. N. Farrell. 2012. "Wind Farm Siting and Protected Areas in Catalonia: Planning Alternatives or Reproducing 'one-dimensional Thinking'?" *Sustainability* 4: 3180–3205. doi:10.3390/su4123180.

Bahro, R. 1977. Die Alternative: Zur Kritik des real existierenden Sozialismus. Frankfurt am Main: Europäische Verlaganstalt.

Barber, B. 1984. *Strong Democracy: Participatory Politics for a New Age*. London: University of California Press.

Blake, A., 2017. "Kellyanne Conway Says Donald Trump´s Team Has 'Alternative Facts.' Which Pretty Much Says it All." *The Washington Post*. Accessed October 2018. https://www.washingtonpost.com/news/the-fix/wp/2017/01/22/kellyanne-conway-says-donald-trumps-team-has-alternate-facts-which-pretty-much-says-it-all/

Blühdorn, I. 2000. *Post-ecologist Politics: Social Theory and the Abdication of the Ecologist Paradigm*. London: Routledge.

Borger, J., and M. Chulov, 2018. "Trump 'Stands With' Saudi Arabia and Defends Crown Prince over Khashoggi." *The Guardian*. online ed. https://www.theguardian.com/us-news/2018/nov/20/trump-saudi-arabia-jamal-khashoggi-crown-prince

Brennan, T. 2004. *The Transmission of Affect*. Ithaca, NY: Cornell University Press.

Camus, A. [1951] 1953. *The Rebel*. London: Hamish Hamilton.

Carson, R. 1963. *Silent Spring*. London: Hamish Hamilton.

Chalmers, A. 2003. "The Theory-Dependence of the Use of Instruments in Science." *Philosophy of Science Association* 70: 493–509. doi:10.1086/376924.

Chapin, J. K., K. A. Moxon, R. S. Markowitz, and M. A. L. Nicolelis. 1999. "Real-time Control of a Robot Arm Using Simultaneously Recorded Neurons in the Motor Cortex." *Nature Neuroscience* 2: 664–670. doi:10.1038/10223.

Cilliers, P. 2005. "Complexity, Deconstruction and Relativism." *Theory, Culture & Society* 22: 255–267. doi:10.1177/0263276405058052.

Couldry, N., A. Fotopoulou, and L. Dickens. 2016. "Real Social Analytics: A Contribution Towards A Phenomenology of A Digital World." *The British Journal of Sociology* 67: 118–137. doi:10.1111/bjos.2016.67.issue-1.

Cózar, A., E. Martí, C. M. Duarte, J. García-de-Lomas, E. van Sebille, T. J. Ballatore, V. M. Eguíluz, et al. 2017. "The Arctic Ocean as a Dead End for Floating Plastics in the North Atlantic Branch of the Thermohaline Circulation." *Science Advances* 3: e1600582. doi:10.1126/sciadv.1600582.

Creasey, G. H., J. H. Grill, M. Korsten, R. Betz, R. Anderson, J. Walter, and I. N. R. Group. 2001. "An Implantable Neuroprosthesis for Restoring Bladder and Bowel Control to Patients with Spinal Cord Injuries: A Multicenter Trial." *Archives of Physical Medicine and Rehabilitation* 82: 1512–1519. doi:10.1053/apmr.2001.25911.

Crutzen, P. J., and E. F. Stoermer. 2000. "The Anthropocene." *IGBP [International Geosphere-Biosphere Programme] Newsletter* 41: 17–18.

De Beukelaer, C., and M. Fredriksson. 2018. "The Political Economy of Intellectual Property Rights: The Paradox of Article 27 Exemplified in Ghana." *Review of African Political Economy* 45: 1–21.

Dennett, D. C. 1992. *Consciousness Explained*. London: Penguin Books.

Dewey, J. [1929] 1958. *Experience and Nature*. New York: Dover Publications.

Diaz, R. J., and R. Rosenberg. 2008. "Spreading Dead Zones and Consequences for Marine Ecosystems." *Science* 321: 926. doi:10.1126/science.1156401.

Díaz, S., S. Demissew, J. Carabias, C. Joly, M. Lonsdale, N. Ash, A. Larigauderie, et al. 2015. "The IPBES Conceptual Framework — Connecting Nature and People." *Current Opinion in Environmental Sustainability* 14: 1–16. doi:10.1016/j.cosust.2014.11.002.

Eagleton, T. 2015a. *Hope without Optimism*. Charlottesville, VA: University of Virginia Press.

Eagleton, T., 2015b. "The Slow Death of the University." *The Chronicle of Higher Education*, April 6.

Farrell, K. 2004. "Recapturing Fugitive Power: Epistemology, Complexity and Democracy." *Local Environment* 9: 469–479. doi:10.1080/1354983042000255360.

Farrell, K. N. 2007. "Living with Living Systems: The Co-evolution of Values and Valuation." *International Journal of Sustainable Development and World Ecology* 14: 14–26. doi:10.1080/13504500709469704.

Farrell, K. N. 2008. "The Politics of Science and Sustainable Development: Marcuse's New Science in the 21st Century." *Capitalism Nature Socialism* 19: 68–83. doi:10.1080/10455750802559616.

Farrell, K. N. 2011a. "The Politics of Science: Has Marcuse's New Science Finally Come of Age?" In *Critical Ecologies: The Frankfurt School and Contemporary Environmental Crises*, edited by A. Biro, 73–107. Toronto: University of Toronto Press.

Farrell, K. N. 2011b. "Snow White and the Wicked Problems of the West: A Look at the Lines between Empirical Description and Normative Prescription." *Science, Technology, and Human Values* 36: 334–361. doi:10.1177/0162243910385796.

Farrell, K. N. 2017. "The Anthropocene and Rudolf Bahro's Megamachine: A Preliminary Exploration of Some Key Relationships between Marcuse's Concept of Technology as Ideology and Bahro's Discourse on the Decomposition of Late Industrial Societies." *Annuaire Roumain d'Anthropologie* 54: 117–132.

Farrell, K. N. [2005] 2009. *Making Good Decisions Well: A Theory of Collective Ecological Management*. Aachen, Germany: Shaker Verlag GmbH.

Farrell, K. N., and D. Barkin, 2016. "Risks of Undisciplined Thinking and Possible Advantages of Epistemological Solidarity: A Comparison of Options." Entitle Conference: Undisciplined Environments, Stockholm, March 20–23.

Farrell, K. N., T. Luzzati, and S. van den Hove. 2013. "What Lies beyond Reductionism? Taking Stock of Inter-disciplinary Research in Ecological Economics." In *Beyond Reductionism: A Passion for Inter-disciplinarity*, edited by K. N. Farrell, T. Luzzati, and S. van den Hove, 36–75. New York: Routledge.

Farrell, K. N., A. Saltelli, S. Tarantola, and D. Gatelli. 2007. *Coping with Corruption within Complex Systems: Methods for Managing Modelling Ambiguity*. Boston, MA: International Conference on Complex Systems.

Fuller, S. 2003. *Kuhn Vs. Popper: The Struggle for the Soul of Science*. Cambridge: Icon Books.

Fuller, S. 2018. *Post-Truth: Knowledge as a Power Game*. London: Anthem Press.

Funtowicz, S. O., and J. R. Ravetz. 1991. "A New Scientific Methodology for Global Environmental Issues." In *Ecological Economics: The Science and Management of Sustainability*, edited by R. Costanza, 137–152. New York: Columbia University Press.

Funtowicz, S. O., and J. R. Ravetz. 1992. "The Good, the True and the Post-modern." *Futures* 24: 963–976. doi:10.1016/0016-3287(92)90131-X.

Funtowicz, S. O., and J. R. Ravetz. 1993. "Science for the Post-Normal Age." *Futures* 25: 739–755. doi:10.1016/0016-3287(93)90022-L.

Funtowicz, S. O., and J. R. Ravetz. 1997. "The Poetry of Thermodynamics - Energy, Entropy/exergy and Quality." *Futures* 29: 791–810. doi:10.1016/S0016-3287(97)00059-1.

Gibbons, M., C. Limoges, H. Nowotny, S. Schwartzman, P. Scott, and M. Trow. 1994. *The New Production of Knowledge. The Dynamics of Science and Research in Contemporary Societies*. London: Sage.

Glannon, W. 2016. "Ethical Issues in Neuroprosthetics." *Journal of Neural Engineering* 13: 021002. doi:10.1088/1741-2560/13/2/021002.

Goethe, J. W. V. [1808–1832] 1938. *Faust. (Erster und Zweiter Teil). Mit einer Einleitung "Faust und die Kunst" von Max von Boehn*. Berlin: Askanischen Verlag.

Gould, S. J. 2003. *The Hedgehog, the Fox and the Magister's Pox*. London: Jonathan Cape.

The Guardian. 2018. "Trump: Khashoggi Case Will Not Stop $110bn US-Saudi Arms Trade – Video." Accessed October 2018. https://www.theguardian.com/us-news/video/2018/oct/12/trump-khashoggi-case-will-not-stop-110bn-us-saudi-arms-trade-video

Habermas, J. 1984. *The Theory of Communicative Action – Vol 1: Reason and the Rationalization of Society*. London: Heinemann.

Haraway, D. 1988. "Situated Knowledges: The Science Question in Feminism and the Privilege of Partial Perspective." *Feminist Studies* 14: 575–599. doi:10.2307/3178066.

Haraway, D. 2016. *Staying with the Trouble: Making Kin in the Chthulucene*. Durham, NC: Duke University Press.

Hohl, K., and M. A. Conway. 2016. "Memory as Evidence: How Normal Features of Victim Memory Lead to the Attrition of Rape Complaints." *Criminology & Criminal Justice* 17: 248–265. doi:10.1177/1748895816668937.

Horkheimer, M., and T. W. Adorno. [1944] 2006. *Dialektik der Aufklärung: Philosophische Fragmente*. Frankfurt am Main: Fischer Taschenbuch Verlag.

Hunt, E., 2017. "Trump's Inauguration Crowd: Sean Spicer's Claims versus the Evidence." *The Guardian*. Accessed October 2018. https://www.theguardian.com/us-news/2017/jan/22/trump-inauguration-crowd-sean-spicers-claims-versus-the-evidence

Ignatius, D., 2018. "Why Was MBS so Afraid of Jamal Khashoggi?" *The Washington Post*. https://www.washingtonpost.com/opinions/global-opinions/let-jamal-khashoggi-be-a-beacon-of-light-even-in-death/2018/10/25/94249cce-d896-11e8-aeb7-ddcad4a0a54e_story.html?utm_term=.5f7eb8fe8d3e

IPBES. 2019. *The IPBES Global Assessment on Biodiversity and Ecosystem Services*. Bonn: IPBES.

IPCC. 2018. "Global Warming of 1.5 °C: An IPCC Special Report on the Impacts of Global Warming of 1.5 °C above Pre-industrial Levels and Related Global Greenhouse Gas Emission Pathways, in the Context of Strengthening the Global Response to the Threat of Climate Change, Sustainable Development, and Efforts to Eradicate Poverty (Summary for Policymakers)." *Intergovernmental Panel on Climate Change (IPCC)*. http://report.ipcc.ch/sr15/pdf/sr15_spm_final.pdf

Kant, I. [1784] 1991. "An Answer to the Question; 'What Is Enlightenment'." In *Kant: Political Writings (Cambridge Texts in the History of Political Thought)*, edited by H. S. Reiss, 54–60. Cambridge: Cambridge University Press.

Kant, I. [1787] 1990. "The Critique of Pure Reason." In *Great Books of the Western World*, edited by M. J. Adler. Vol. 39. London: Encyclopædia Britannica.

Kates, R. W., W. C. Clark, R. Corell, M. J. Hall, C. C. Jaeger, I. Lowe, J. J. McCarthy, et al., 2001. "Sustainability Science: Research and Assessment Systems for Sustainability Environment and Natural Resources Program John F. Kennedy School of Government." *Faculty Research Working Paper Series 2000-33*. Boston: Harvard University.

Kuhn, T. S. [1962] 1970. *The Structure of Scientific Revolutions*. 2nd ed. London: University of Chicago Press.

Lander, E. 2000. "Ciencias sociales: saberes coloniales y eurocéntricos." In *La colonialidad del saber: eurocentrismo y ciencias sociales Perspectivas latinoamericanas*, edited by E. Lander, 11–40. Buenos Aires: CLACSO.

Latour, B. 2004. *Politics of Nature: How to Bring the Science into Democracy*. Cambridge, MA: Harvard University Press.

Latour, B., and S. Woolgar. 1979. *Laboratory Life: The Social Construction of Scientific Facts*. London: Sage.

Machin, A. 2019. "Democracy and Agonism in the Anthropocene: The Challenge of Knowledge, Time and Boundary." *Environmental Values* 28: 347–365. doi:10.3197/096327119X15519764179836.

Malm, A., and A. Hornborg. 2014. "The Geology of Mankind? A Critique of the Anthropocene Narrative." *The Anthropocene Review* 1: 62–69. doi:10.1177/2053019613516291.

Marcuse, H. 1969. *An Essay on Liberation*. London: Allen Lane The Penguin Press.

Marcuse, H. 1978. "Protosozialismus und Spätkapitalismus – Versuch einer revolutionstheoretischen Synthese von Bahros Anzatz." *Kritik* 19 (6): 5-27.

Marcuse, H. [1955] 1972. *Eros and Civilisation*. London: Abacus.

Marcuse, H. [1964] 1991. *One-Dimensional Man: Studies in the Ideology of Advanced Industrial Society*. London: Routledge.

Mayeda, A., 2018. "Technology: IMF Says the Global Smartphone Boom Has Reached Its Peak." *bloomberg.com. Bloomberg*. Accessed October 2018. https://www.bloomberg.com/news/articles/2018-04-17/imf-says-the-global-smartphone-boom-has-reached-its-peak

McCain, J. S. 1974. *Individual Research Project: The Code of Conduct and the Vietnam Prisoners of War*. Washington, D.C: National War College.

Merchant, C. 1981. *The Death of Nature: Women, Ecology, and the Scientific Revolution*. New York: Harper.

Merchant, C. 2006. "The Scientific Revolution and the Death of Nature." *Isis* 97: 513–533. doi:10.1086/508090.

Merchant, C. 2008. "'The Violence of Impediments': Francis Bacon and the Origins of Experimentation." *Isis* 99: 731–760. doi:10.1086/597767.

Mignolo, W. D. 2000. "La colonialidad a lo largo y a lo ancho: el hemisferio occidental en el horizonte colonial de la modernidad." In *La Colonialidad Del Saber: Eurocentrismo y Ciencias Sociales Perspectivas Latinoamericanas*, edited by E. Lander, 55–86. Buenos Aires: CLACSO.

Mirowski, P. 2011. *Science-Mart: Privatizing American Science*. Cambridge, MA: Harvard University Press.

Moore, J. 2016. *Anthropocene or Capitalocene? Nature, History, and the Crisis of Capitalism*. Oakland, CA: PM Press.

Mouffe, C. 2005. *On the Political*. London: Verso.

Murdock, E. G. 2018. "Unsettling Reconciliation: Decolonial Methods for Transforming Social-Ecological Systems." *Environmental Values* 27: 513–533. doi:10.3197/096327118X15321668325948.

Nietzsche, F. W. 1882. "Die fröhliche Wissenschaft." *The Gutenberg Project*. Accessed October 2018. http://gutenberg.spiegel.de/buch/die-frohliche-wissenschaft-3245/1

O'Neill, J. 1993. *Ecology, Policy and Politics: Human Well-Being and the Natural World*. London: Routledge.

Pellizzoni, L. 2003. "Uncertainty and Participatory Democracy." *Environmental Values* 12: 195–224. doi:10.3197/096327103129341298.

Popper, K. [1959] 1968. *The Logic of Scientific Discovery*. New York: Harper & Row Publishers.

Prigogine, I., and I. Stengers. 1984. *Order Out of Chaos: Man's New Dialogue with Nature*. London: Heinemann.

Ravetz, J. R. 1971. *Scientific Knowledge and Its Social Problems*. Oxford: Clarendon Press.

Ravetz, J. R. 1990. *The Merger of Knowledge with Power: Essays in Critical Science*. London: Mansell.

Ravetz, J. R. 2013. "The Maturing of the Structural Contradictions of Modern European Science: An Exploratory Sketch." In *Beyond Reductionism: A Passion for Inter-disciplinarity*, edited by K. N. Farrell, T. Luzzati, and S. van den Hove, 76–89. New York: Routledge.

Rittel, H. W. J., and M. M. Webber. 1973. "Dilemmas in a General Theory of Planning." *Policy Sciences* 4: 155–169. doi:10.1007/BF01405730.

Saltelli, A. 2018. "Why Science's Crisis Should Not Become a Political Battling Ground." *Futures* 104: 85–90. doi:10.1016/j.futures.2018.07.006.

Schmidtko, S., L. Stramma, and M. Visbeck. 2017. "Decline in Global Oceanic Oxygen Content during the past Five Decades." *Nature* 542: 335. doi:10.1038/nature21399.

Shiva, V. 1989. *Staying Alive: Women, Ecology and Development*. London: Zed Books.

Solinsky, R., and L. Specker Sullivan. 2018. "Ethical Issues Surrounding a New Generation of Neuroprostheses for Patients with Spinal Cord Injuries." *PM&R* 10: S244–S248. doi:10.1016/j.pmrj.2018.05.020.

Steffen, W., J. Rockström, K. Richardson, T. M. Lenton, C. Folke, D. Liverman, C. P. Summerhayes, et al. 2018. "Trajectories of the Earth System in the Anthropocene." *Proceedings of the National Academy of Sciences* 115: 8252. doi:10.1073/pnas.1810141115.

Stengers, I. 1997. *Power and Invention: Situating Science*. London: University of Minnesota Press.

Topping, C. J., H. Fjelsted Alrøe, K. N. Farrell, and V. Grimm. 2015. "Per Aspera Ad Astra: Through Complex Population Modelling to Predictive Theory." *The American Naturalist* 186: 669–674. doi:10.1086/683181.

Toulmin, S. 1990. *Cosmopolis: The Hidden Agenda of Modernity*. New York: Free Press.

Wakefield, N. 1990. *Postmodernism: The Twilight of the Real*. London: Pluto Press.

Warwick, K., M. Gasson, B. Hutt, I. Goodhew, P. Kyberd, B. Andrews, P. Teddy, and A. Shad. 2003. "The Application of Implant Technology for Cybernetic Systems." *Archives of Neurology* 60: 1369–1373. doi:10.1001/archneur.60.10.1369.

Waters, C. N., J. Zalasiewicz, C. Summerhayes, A. D. Barnosky, C. Poirier, A. Gałuszka, A. Cearreta, et al. 2016. "The Anthropocene Is Functionally and Stratigraphically Distinct from the Holocene." *Science* 351: aad2622–aad2622. doi:10.1126/science.aad2622.

Weber, M. [1904] 2015. *Die protestantische Ethik und der Geist des Kapitalismus: Neuausgabe der ersten Fassung von 1904-05 mit einem Verzeichnis der wichtigsten Zusätze der Sozialwissenschaften (German Edition)*. Heidelberg: Springer.

White House, T. 2018. "Statement from President Donald J. Trump on Standing with Saudi Arabia." *The White House*. https://www.whitehouse.gov/briefings-statements/statement-president-donald-j-trump-standing-saudi-arabia/

Woltersdorff, V. A. L. L. 2003. "Queer Theory und Queer Politics." *UTOPIE Kreativ, H* 156: 914–923.

Zalasiewicz, J., M. Williams, A. Haywood, and M. Ellis. 2011. "The Anthropocene: A New Epoch of Geological Time?" *Philosophical Transactions of the Royal Society A* 369: 835–841. doi:10.1098/rsta.2010.0339.

Zengerle, P. 2018. "Top Senators Briefed by CIA Blame Saudi Prince for Khashoggi Death." *Reuters*. https://www.reuters.com/article/us-saudi-khashoggi-cia/top-senators-briefed-by-cia-blame-saudi-prince-for-khashoggi-death-idUSKBN1O32BR

Designing a Transformative Epistemology of the Problematic: A Perspective for Transdisciplinary Sustainability Science

Esther Meyer and Daniela Peukert

ABSTRACT
This paper elaborates on the question of how to design an epistemological foundation for problem-oriented, collaborative forms of research, such as transdisciplinary sustainability research. It picks up approaches of twentieth-century European philosophy to the concept of the problematic and design research. The problematic is explained as a historical epistemological effort. Design research shows parallels to the epistemological thinking of the problematic by contributing to a differentiation and historicity of knowledge and knowledge production itself. Designing is constituted by a nexus of conceptual thinking and creative making, and so designs are drafts themselves. We interweave the thinking of the problematic with the practice of designing in order to open an epistemological perspective in and for transdisciplinary sustainability research. We call this a 'thinking practice of problematic designing,' which describes an epistemological tool as well as a transformative process. Problematic designing is characterized by always being in the making – its designs can grow beyond their conditions of production. By opening up manifold dimensions of transformation, this epistemological approach is oriented towards complexity, enabling the generation of sound and future-relevant knowledge.

Introduction

> Addressing our growing planetary crisis, and attendant symptoms of human and human-ecological disconnect, requires a profound epistemological reorientation regarding how societal structures are conceived and articulated. (Williams et al. 2017, 41)

With these words, Lewis Williams and his colleagues introduce their ideas on a 'Global De-colonial Praxis of Sustainability.' We follow the requirement of an 'epistemological reorientation' (Williams et al. 2017) as formulated in the quote as well as by other authors (Krohn, Grunwald, and Ukowitz 2017; Nowotny 2006). We revisit the question of how *transdisciplinary research* might be equipped with a *transformative epistemological foundation* to face a planet in crisis mode. Transdisciplinary research forms part of sustainability sciences and is in this setting a form of boundary-crossing research (Vilsmaier 2018),[1] oriented towards goals of sustainability, which has emerged from German-speaking Europe for almost 20 years.[2]

Transdisciplinary sustainability research is often realized in projects that focus on specific problems for which a solution is to be found. We begin by referring to a previous study on the term *problem* in European transdisciplinary sustainability sciences (Meyer forthcoming b).[3] According to the results, the term problem on the one hand derives its meaning from a normative orientation towards problem-solving, and on the other hand analytically relates to the process level of research. By focusing on problem-solving, an instrumental (Welch 2011) controllability (Doucet and Janssens 2011) and manageability of knowledge or the research process is supposed. Furthermore, a problem solving orientation 'assumes that the world is a problematic that needs to be solved. [...] The narrow focus required for this approach deals with problems in isolation, as a 'closed system,' and deals with neither context nor relationships' (Haley 2011, 24). The results of the study further disclose that the term problem has neither been explained theoretically nor connected to epistemological issues in the discourse. At the same time, authors highlight exactly this connection and see a need for an elaboration of epistemological implications in transdisciplinary research (Krohn, Grunwald, and Ukowitz 2017; Nowotny 2006).

Welch traces two discourses in his genealogy of interdisciplinarity: 'the critical and instrumental modes of interdisciplinarity' (Welch 2011, 4) that 'are not mutually exclusive' (Welch 2011, 4). Instead, 'their synthetic relationship is essential for developing an interdisciplinary theory as an emergent epistemological innovation' (Welch 2011, 4).

Therefore, we pick up on the critical, philosophical concept of *the problematic* (Hörl and Leistert 2019; Meyer forthcoming a) that has been constituted in the 'breakdown of traditional epistemological structures' (Welch 2011, 4). We then establish a link to *design*, as a pragmatic and methodology-oriented (Repko and Szostak 2008) product of the instrumental strand, in order to inform transdisciplinary sustainability research. Our goal is to open an epistemological perspective in and for transdisciplinary sustainability research that does not rely on a mechanistic solvability of problems, nor creates epistemic problems to control the future of science. Our thinking is based on the assumption that collaborative research forms in themselves pave the way for science and research in planetary crisis mode, if they are 'navigating transformative TDR [transdisciplinary research] processes in and under fluid social conditions' (Van Breda and Swilling 2019, 826). For this kind of collaboration, however, we do consider the need to unlearn the language of control and set up new alphabets,[4] 'practice theories [...] that [...] attempt to provide a new vocabulary to describe the world' (Nicolini 2013, 9 in Maguire 2018, 113).

In this paper, we explain the philosophical concept of the *problematic*,[5] which has been developed in a twentieth-century French epistemological tradition. The concept of the problematic emphazises French theory's differentiation of the subject–object dichotomy (Maniglier 2012), which makes it interesting for transdisciplinary research (Nicolescu 2010), where research collectives go beyond the traditional divide of an active researcher and a passive object of research (Vilsmaier, Brandner, and Engbers 2017). Thus, the concept of the problematic contributes to our goal of epistemologically substantiating the discourse. Transdisciplinary sustainability research constitutes a 'profound turn in Western thought' (Welch 2011, 4), and thus carries a historical share of thinking the problematic. Nevertheless, the discourse that we are looking at builds on patterns of separation, such as the distinction of problem and solution (Nicolescu 2010). Subsequently, we propose *designing* as a creative practice that offers a pathway towards a transformative epistemology, in which thinking and doing are strongly interconnected. The reason we bring together the philosophy of the problematic and design lies in the twofold nature of design. On the one hand, design reveals its strong epistemological force in the process of designing drafts as (manifested) ideas. On the other hand, designing as a practice and process produces tangible artefacts that are able to visualize and concretize transformations. By bringing together problematic thinking and design, we use the problematic as a philosophical base and interweave it with design to develop an epistemological approach that is able to unfold intended connections to transformative practice. We condense distinctive characteristics from the examination of the problematic and put them into relation with design to develop the thinking practice of *problematic designing*. By thinking practice, we mean the thinking that influences practice through being informed by practice. The basic idea of problematic designing is to embed the practice of designing in the epistemology of the problematic. We understand the *problematic* in the sense of a designed draft. For us, it 'is the beginning of a thought process that moves towards a consciousness of a way of being as well as a way of doing in the world which is [...] always a process of becoming' (Maguire 2018, 110) and it 'is more about an attitude to knowledge than it is about knowledge; [...] it is generative not replicative' (Maguire 2018, 114).

Afterwards, we discuss the perspective of problematic designing in relation to transdisciplinary sustainability research with regards to two of its main characteristics (Engbers 2018): heterogeneous actor relationships and solving complex problems. In combining these two characteristics we propose a *transformative epistemology of problematic designing* as a new perspective for transdisciplinary sustainability research, based on epistemological breaks, as an alternative primer to a dilemma of control. In the transformative epistemology of problematic designing we see a possibility of the (re-)politicization, and thus (re-)futurization (Raven and Elahi 2015, 51), of sustainability-related issues.

Problematic designing serves as a thought process as well as a research and design process. In this conceptual paper, we will primarily focus on the thought process – practical implications for transdisciplinary and transformative research and design processes will be addressed in upcoming research.[6]

The Problematic

We reflect upon the problematic as described by French and Belgian philosophers over the last century. For this purpose, authors and concepts are introduced from this philosophical direction in the current chapter. We aim to historically understand the concept of problematic thinking in order to work out the transformative epistemology of problematic designing for transdisciplinary sustainability research. We begin to derive an epistemological problematic from Gaston Bachelard,[7] continue with onto-epistemological stances that we explain through the study of literature on and by Gilbert Simondon,[8] and finally discuss (research)-ethical considerations by Isabelle Stengers.[9]

With Bachelard, the problematic marks an *epistemological break* (Maniglier 2012, 22) in the understanding of a demarcation between subject and object.[10] The problematic can be understood as an 'applied doubt',[11] 'a doubt specified by the object of knowledge' (Bachelard 2012, 30). It is in the sense of a doubt arising from a concrete situation against the background of universal assumptions. This doubt arises always and necessarily[12] and marks the knowledge-generating moment. Through a problematic epistemology everything, including knowledge, is constituted in correlation,[13] as Patrice Maniglier writes: 'Neither objects nor subjects, neither things nor minds, exist primarily; there are only problems, which institute the very possibility of the correlation' (Maniglier 2012, 22). Correlation thus has an epistemological function. Problems as epistemic starting points therefore 'constitute that which makes it important, relevant, critical, to know about' (Maniglier 2012, 21) an object of knowledge. An object is always preceded by an objectification. In the epistemology of the problematic, 'the bases for knowledge are themselves put to the test, and brought into question by the question' (Maniglier 2012, 22) of the frame of knowledge production itself. Bachelard is concerned with the epistemological, as a history of science[14] itself, which for him breaks with ontological fundamentalism. 'It is noteworthy that at a time when others were writing about the unity of science and of the scientific method, Bachelard was talking about the plurality of regional rationalisms' (Tiles 2012, 26). To summarize, the problematic works as a theorization on the mediation between the scientific subject and the object, and on the knowledge-generating moment. For Bachelard, it is a component with which he explains his historical epistemology.

An onto-epistemological stance towards the problematic is expressed by Gilbert Simondon. For him, the problematic is an auxiliary construction in the explanation of *becoming an individual* – what he calls individuation of living beings. Individuation means that the living can only become in relation to its condition (Scott 2014). Individuation takes place in a situated, open process, in which the problematic is also described in a relational sense (Voss 2018). Simondon thus calls it *problematic becoming*, which is characterized by inventing axiomatics. Axiomatics he explains as internal structures that make it possible to live in relation to changing conditions (Voss 2018). '[T]he social dimension of the transindividual' (Voss 2018, 96) arises out of this problematic movement. It forms both the living individual and their conditions while also being formed by them (Voss 2018, 96). They keep each other moving. Such continuously moving, living structures Simondon calls *milieu* (Voss 2018, 96). Individuation may crystallize in a *metastable* – structurally compatible – state in the 'association of individual and milieu' (Voss 2018, 99; Scott 2014, 4). The problematic has the function of inventing this association into ever new, metastable states, which itself is the function of psychic life (Voss 2018). With regard to problem solving, Simondon writes,

> [t]o solve a problem is to be able to step over it, to be capable of recasting the forms that are given within the problem and in which it consists. The solution of real problems is a vital function presupposing a recurrent mode of action that cannot exist in the machine: the recurrence of the future with respect to the present, of the virtual with respect to the actual. (Simondon 2017, 156)

Simondon adds an unconditional epistemological moment to this rather ontological level of description, which we will have to examine in more detail in this work: Individuation always takes place in particular, concrete cases that need to be explored (Voss 2018) and that marks the knowledge-generating – problematic – moment.

Research-ethical considerations are introduced by Isabelle Stenger's concept of *ecology of practices*, which, as a tool for thinking (Stengers 2005), is connected to problematic thinking. She continues in her concept with the particular, situated and plural thinking that is incompatible with universal or totalitarian epistemologies and problems, believing in the truth (Fichant 1975; Harrasser and Solhdju 2016; Stengers 2005). Regarding the term problem she formulates, 'a problem is always a practical problem, never a universal problem mattering for everybody. Problems of the ecology of practices are also practical problems in this strong sense, that is problems for practitioners' (Stengers 2005, 193). Ethical assessment of research outcomes takes place within an ecology of practices, based on effectiveness, consequences and matter, by those who affect the practice (Harrasser and Solhdju 2016; Stengers 2005), and in a 'specific milieu' (Harrasser and Solhdju 2016, 74, own translation). Practice and surrounding environment condition each other and emerge as milieu-like ecology. Similar to Simondon, the epistemological moment lies in the diagnosis of the 'relationship of relevance' between (Stengers 2005, 185) the material and symbolic relations, with and within the milieu. For Stengers, a milieu is essentially characterized by entanglement between (epistemic) practice, which includes the practitioners, and its effects. The effects become new surroundings for practices (Stengers 2005), in which she sees the ecological moment of a practice. She regards an ecology of practices as good and equivalent to modern modes of cognition (Harrasser and Solhdju 2016) around 'rational universality' (Stengers 2005, 196). An ecology of practices also realizes ethics, in the form of ethical experimentation (Harrasser and Solhdju 2016; Stengers 2010). Practice and surrounding environment condition each other, and emerge as milieu-like ecology – as a moment of ethical knowledge generation (Stengers 2005). Stengers incorporates the structure of the problematic in her ecology of practices as a transformative tool for thinking (Harrasser and Solhdju 2016) within her ecological epistemology. The justification modes for a problem and forms of research to solve it shift from justification based on objectivity and universality to justification based on the adaptation to a socio-ecological situation.

In summarizing this chapter, we will formulate a kind of working definition of the problematic by contrasting and embedding the components above. Epistemology has been constituted in modern philosophy by a separation between subject and object (Fichant 1975). Bachelard's epistemology marks an epistemological break (Fichant 1975). Epistemological break means a 'new concept of epistemology' (Fichant 1975), for which he introduces the structure of the problematic as an element of theory that builds on the mediation between the scientific subject and the object. Through this epistemological break, epistemology becomes historical. Simondon can build on this and develops his onto-epistemology, which is based on particular individual–milieu connections. Here, similar to Bachelard, the thinking structure of the problematic takes on an epistemological function for the living. Stengers, in her ecology of practices, incorporates the thinking structure of the problematic in its epistemological function. It is a transformative tool for thinking (Harrasser and Solhdju 2016) characterized by many epistemological breaks.

With the concept of the problematic, French theory introduced a historical philosophy of science that was able to work out the dominant notions of separation: a differentiation of the subject–object dichotomy, the external and internal, world–psyche relations, or the individual and the collective. It reminds us of the historicity of how knowledge is generated and explained as knowledge. The question of how we know that we know is historically, socio-culturally, economically and politically situated. The problematic is to be understood as epistemologically relevant insofar as epistemology itself is to be understood in the following meaning: 'different ways of knowing, acting and being' (Vilsmaier, Brandner, and Engbers 2017, 170) are made up of certain cultural traditions, societal paradigms and geopolitical discourses and their historical contingencies. So the problematic is not about defining or questioning what the problem is, but how knowledge is generated (Wulz 2014, 70). Thus, epistemology has a problematic and essential function in questioning knowledge according to its respective validity. It therefore introduces the possibility of epistemologically breaking with possibly dominant normative notions of science and methods of knowledge generation, and is finally understood as an ecology of practices (Stengers 2005).

In the following section, we will interweave the thinking of the problematic with the practice of designing. In doing so, we develop the thinking practice of *problematic designing* to demonstrate its transformative power as an epistemology for transdisciplinary sustainability research.

Problematic Designing

In order to open up the historical epistemology of the problematic for transdisciplinary sustainability research, we connect it to the practice of designing and its speculative drafts in order to be able to generate sound and future-relevant knowledge. Designing is understood as a process that produces designs, which have the character of manifested ideas, always containing elements of future, uncertainty and provisionality, and can therefore be called speculative drafts. We understand future-relevant knowledge to be generated always in and with practice. Thereby, we follow a thought that has already been established by other authors who use the term 'design as a pragmatic orientation for the selection of knowledge' (Krohn, Grunwald, and Ukowitz 2017, 342, own translation) within transdisciplinary processes, and develop it further epistemologically.

Designing is considered a planning action that transforms 'existing situations into preferred ones' (Simon 1969) by using creative methods for visualization. It moves between two poles that Banz (2016) describes as 'design doing' and 'design thinking.' 'Design doing' primarily describes the making, which means the output-oriented tangible practice of design, whereas 'design thinking' includes the cognitive planning aspects of design. As a result, design is no longer solely focused on artefacts, with its aesthetic and functional demands that go hand in hand with shaping forms, but on the transformative processes of planning and problem solving, which can also be applied in political or social processes (Escobar 2018; Fry 2011; Manzini 2015; Papanek 1971).

The specific epistemic quality of designing lies in a nexus of conceptual thinking and creative making that can result in visual and tangible objects leading to new knowledge and transformative power as they change situations by their design (Peukert and Vilsmaier 2019). Designed objects have a closed nature, whereas a design process itself has a procedural and open character. This reveals the dialectic nature of design. As there is no end to a design process and no final solution, designing results in a forward-oriented openness. We will never know under which conditions a design will be consulted or assessed in the future. As Maniglier points out in his thinking of the problematic, the solution is not just given, but 'in the making' in the form of a metastable state similar to the provisional final phase of designs: 'if we problematize the world [...] [b]ecause our own thought proceeds as a process that structures a set of propositions. The structure is neither given in advance nor constructed: it is all in the making' (Maniglier 2012, 23). Design theories show historical parallels to the emergence of the epistemological thinking of the problematic by contributing to a differentiation, historicity and situatedness of knowledge and knowledge production itself (Dorst 2003): the practice of designing is epistemologically situated, which means it is constituted by a specific epistemic quality, consisting of a nexus of conceptual thinking and creative making, as do designs themselves.

After referencing design and designing, and connecting it to thinking the problematic, we will introduce and explain the thinking practice of problematic designing as an epistemological tool. Therefore, we have visualized a two-dimensional representation of the epistemology of problematic designing (see Figure 1). This is influenced by Simondon´s concept of time: 'time itself is essence, not as development starting from an origin or tendency towards some end, but rather as resolute constitution of being' (Simondon in Scott 2014, 6). In his understanding time is not determined by an origin or an end, but is constituted by being (Figure 1).

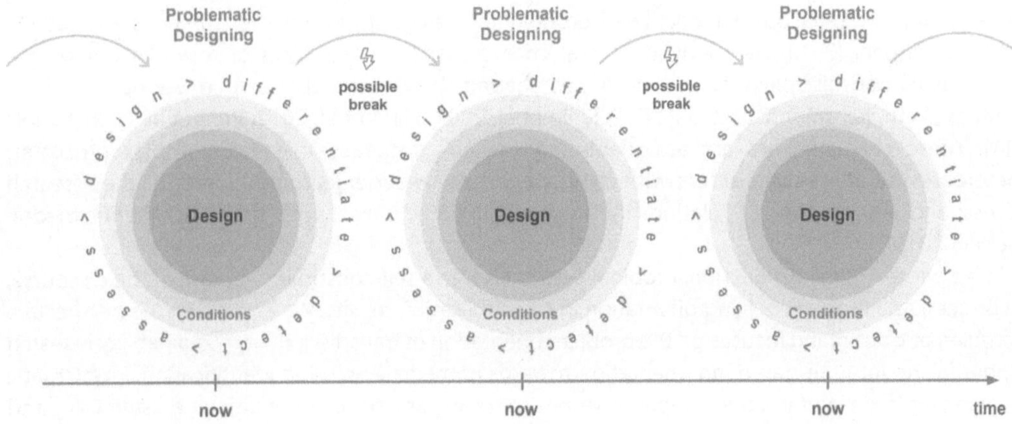

Figure 1. Epistemology of problematic designing.

The timeline in Figure 1 represents different stages of designs always in the present. These are vibrant designs embedded in milieu-like, different environmental conditions. The problematic is the differentiation between the design and the conditions that generate it, and that in turn can be generated by the design. It is a reflexive feedback movement between the initial situation, including the determining normative goal, and the result. The following steps describe problematic designing: differentiate, detect, assess (imbalance) and design. Here we recognize a proximity to the philosophical pragmatism of John Dewey, where 'reflection is provoked when action is faced with obstacles of some sort' (Barnett 2015, 21). Clive Barnett further writes, '[a]fter Dewey, problematic situations are not understood as sudden intrusions into a settled environment, but rather as something becoming off kilter in the course of ongoing activity' (Barnett 2015, 22). In the iterative movements of actively becoming, described in the quote, problematic designing focuses on problematizing and designing on an equal footing.

In the four steps (see Figure 1) lie the *transformative moment* of problematic designing. The transformative moment moves in a complementary way between forms of adaptation of the designs to existing conditions or the complete break with these conditions. 'The historicity of the sciences thus has an open future ahead of them – their objects are characterized by never being finished' (Wulz 2014, 70, own translation). Problematic designing is open-ended. It can question the respective validity of research, knowledge and knowledge production. Furthermore, the results can challenge possibly dominating, normative notions of science and methods of knowledge generation or produce an epistemological break, even if we cannot think it.[15] To regard these possible breaks as epistemologically essential and inherently in the living is a quality of problematic designing, which makes it interesting for sustainability research. The epistemology of the problematic is extended by Simondon's terminology of *being* and *becoming*. He says in his *Psychic and Collective Individuation*, here requoted by David Scott, 'What happens occurs in the form of a putting into question of being, in other words, in the form of the element of an open problematic' (Scott 2014, 6). Being inevitably brings up questions of cognition, which is the moment of understanding and mediating with the environment.[16] Accordingly, problematic designing always happens now, as the radical now is the coordinate point of the living being. The question arises: How can one be detached from the conditions of the design while designing, to do justice to the open-ended process that also enables epistemological breaks? The constant iteration of the four steps of problematic designing allows this apparent contradiction, as in each step every aspect of the conditions can be called into question.

Discussing Implications for Transdisciplinary Sustainability Research

In this section, we refer our perspective of problematic designing to transdisciplinary sustainability research. Engbers (2018) has derived several characteristics of this form of research out of five definitions of transdisciplinarity that decisively shaped the current discourse in German-speaking countries (Jahn, Bergmann, and Keil 2012; Klein et al. 2001; Lang et al. 2012; Pohl et al. 2008; Scholz 2011). These characteristics are an orientation towards societally relevant, complex problems; a heterogeneity of involved actors with their specific knowledge; learning as part of the research process; and a strengthened relationship amongst involved actors with their specific knowledge (Engbers 2018).

As we wrote in the introduction, problem orientation, as a self-constituent feature of the discourse, can be seen as a problem-solving orientation in the publication results. We argue that the perspective of problematic designing focuses on the problem orientation of transdisciplinary sustainability research without falling into the trap of an orientation towards immediate solution identification. Problematic designing instead maintains the complex relations to society as part of its conditions, epistemically and epistemologically, as Van Breda and Swilling (2019) and Welch (2011) have also emphasized. We understand the *problematic* in the sense of a designed draft within the thinking practice of problematic designing allowing, as Isabell Schrickel writes in reference to Gilles Deleuze, 'an intrinsic differential evolution of a problematic situation' (Schrickel forthcoming). Instead of a one-dimensional, solution-oriented directionality to eliminate problems (Schrickel forthcoming) that have been identified in the past, a variety of *dimensions of transformation* are inherent in a design. They may vary between adaptation to epistemic conditions and epistemological break(see Figure 1) – which extends the horizon of a description of knowledge generation to a change in the generation of knowledge itself. By opening up manifold dimensions of transformation, the epistemological approach of problematic designing is oriented towards complexity, with its simultaneities and multiple levels, towards a change that changes change. We will thus call it a *transformative epistemology of the problematic*, and suggest that problematic designing opens new perspectives for transdisciplinary sustainability research. Transdisciplinary sustainability research is a collaborative form of research that brings different epistemologies into play – therefore it also generates manifold epistemological breaks. The fundamental consideration of this epistemological level may be the basis for a 'heterogeneity of specific actors with their specific knowledge' (Engbers 2018), and thus be essential for this kind of research (Krohn, Grunwald, and Ukowitz 2017; Mitchell, Cordell and Fam 2015). In other words, transdisciplinary sustainability research defined by its characteristics is in itself an epistemological break that represents a core theoretical element of collaborative forms of research (Tejedor and Segalas 2018; Welch 2011).[17]

Now one may wonder where the goal orientation of sustainable development is contained in the transformative epistemology of problematic designing? From our perspective, we see a possibility of the (re-)politicization, and thus (re-)futurization (Raven and Elahi 2015, 51), of sustainability-related issues. In openness, with regard to the manifold dimensions of transformation, lies the potential for political work in and through transdisciplinary research projects. By political work we mean work on interests and epistemologies that are based on experiences, and in which nothing matters except the moment of the differentiating encounter of these differences (Maguire 2018; Vilsmaier, Brandner, and Engbers 2017). This work also relates to different interests and epistemological understandings, and thus the future of transdisciplinary (sustainability) research itself; for example, the intercultural exploration of what is understood by transdisciplinarity.

Problematic designing allows the methodological capture of the moment of the problematic difference. Regarding sustainability, this approach results in the possibility of breaking with unsustainable path dependencies that control futures. By thinking the problematic and having the possibility of an epistemological break, problematic designing allows us to leave the taken path and thus for an open 'futurity' (Raven and Elahi 2015) to arise. Unsustainable path dependencies reveal themselves in a post-positivist moment of future orientation, expressed by an accelerated escape to the front towards productivity or a Sisyphus-like problem-solving activity. Instead, problematic designing allows futurity to arise by looking at condition and design at the same time: epistemic conditions do not control the design, nor does the design control the conditions. Transferred to transdisciplinary research, knowledge can only be viewed in the tension between conditions of knowledge generation and its results.

Conclusion

Transdisciplinary sustainability research, as it is constituting itself in Europe, is characterized by a 'heterogeneity of involved actors with their specific knowledge' (Engbers 2018) as well as by the normative goal of problem solving (Nicolescu 2010). Our contribution seeks to provide an epistemological foundation for this form of collaborative research that draws a line between its two characteristics. In our epistemological elaboration, we link two terms. On the one hand, we refer to the *problematic*, which has been constituted in the twentieth-century French philosophical theories. This concept goes beyond a solution-oriented understanding of objectively given problems. On the other hand, we propose *design* as a creative and planning practice, embodying process as well as product. Design has both a reflexive, open character as well as an enclosed nature, which is based on its object characteristic. We develop the epistemological approach of *problematic designing* based on the elaboration of the characteristics of the problematic and design and by relating them to each other. The following epistemic qualities are attributed to problematic designing: it is tied back to design practices as well as their specific historical, socio-cultural, economic and political situatedness. It is an epistemological process inherent in the living, as a permanent state of being in the making, and its transformative moment lies in differentiating the design from its epistemic conditions. We describe this process in four analytical steps: differentiation, detection, assessment and design. Problematic designing forms an open structure as designing is embedded in reflexive movements and constituted by epistemological breaks in a way that design results can grow beyond their conditions.

The thinking practice of problematic designing focuses on the problem orientation of transdisciplinary sustainability research by making transdisciplinary research part of the problem, and thus of complex societal relations and their epistemological foundations (Van Breda and Swilling 2019). Through complex actor relationships in transdisciplinary sustainability research different epistemologies are brought into play. This is why this form of research generates manifold epistemological breaks, simultaneously represents such breaks, and thus extends the horizon of a description of knowledge generation to a change in the generation of knowledge itself. We call this epistemological approach a *transformative epistemology of problematic designing*.

Notes

1. Transdisciplinary sustainability research positions itself as a mode of research between academic and expertise knowledge production and between distinct scientific and societal institutions or organizations. In the course of this, it is complementary to disciplinary and interdisciplinary approaches.

2. Moritz Engbers (2018) identified the current discourse in German-speaking countries based on the following authors: Jahn, Bergmann, and Keil (2012); Klein et al. (2001); Lang et al. (2012); Pohl et al. (2008); Scholz (2011).
3. An exemplary analysis of the term 'problem(s)' has been conducted through English-language article publications in the journal *GAiA*, with the help of computer-assisted discourse studies.
4. See https://www.hkw.de/en/programm/projekte/2019/das_neue_alphabet/das_neue_alphabet_start.php.
5. To avoid misunderstandings: we do not refer to the term used by the Club of Rome, as it was prominently taken up by Manfred Max-Neef (2005).
6. For example, connecting problematic designing to John van Breda's and Mark Swilling's *emergent transdisciplinary design research (ETDR)*, understood as a methodological approach, derived from 'a case study in the informal settlement (slum) of Enkanini in Stellenbosch, South Africa' (Van Breda and Swilling 2019, 823).
7. 1884–1962.
8. 1924–1889.
9. Born in 1949.
10. For 'the concept of problematique initiates a critique of the subject–object relation in the explanation of thought in general and of science in particular. To think is not to try to tell the truth about any particular given objects (be this living organisms, things in motion or brains), as if there was a world out there waiting for us to lay our eyes on it' (Maniglier 2012).
11. Further elaborated as *singular problems* by Patrice Maniglier (2012).
12. Just as the individuation principle and the individuation process necessarily exist incessantly in life (Simondon 2007, 31).
13. Correlationism is a philosophical position claiming that 'a thought of reality that is independent from human knowledge' does not exist (Savransky 2016, 196).
14. 'Every experiment on the reality already informed by science is at the same time an experiment on scientific thought' (Bachelard 2012, 29; Fichant 1975).
15. The quality of an epistemological break, in its radical nature, makes it impossible to think of.
16. We attribute cognition to humans, but would not exclude that it is also possible for other animals.
17. The contentual qualities of the break or the transformation of science differ from those at the time of Bachelard's work. We are aware of it but do not address the differences in this paper.

Acknowledgments

We thank the reviewers of this book, the team of 'Complexity or Control? Paradigms for Sustainable Development,' the team of 'Leverage Points for Sustainability Transformations,' the TD Methods group at the Methodology Center of Leuphana University, Lüneburg, Claudia Mareis, Dena Fam and Bianca Vienni for their support.

Disclosure statement

No potential conflict of interest was reported by the authors.

Funding

This research has been funded by the 'Volkswagenstiftung' under the program 'Science and Scholarship for Sustainable Development.'

References

Bachelard, G. 2012. "Corrationalism and the Problematic." *Radical Philosophy* 173: 27–32.

Banz, C. 2016. "Zwischen Widerstand und Affirmation. Zur wachsenden Verzahnung von Design und Politik." In *Social Design: Gestalten für die Transformation der Gesellschaft*, edited by C. Banz, 11–26. Bielefeld. Transcript.

Barnett, C. 2015. "On Problematization Elaborations on a Theme in 'Late Foucault'." *nonsite.org*, 16. https://nonsite.org/article/on-problematization

Dorst, K. 2003. "The Problem of Design Problems." In: *Expertise in Design: Design Thinking Research Symposium*, edited by E. Edmonds and N. Cross. Sydney: Creativity and Cognition Studios Press.

Doucet, I., and N. Janssens. 2011. "Editorial: Transdisciplinarity, the Hybridisation of Knowledge Production and Space-Related Research." In *Transdisciplinary Knowledge Production in Architecture and Urbanism*, edited by I. Doucet and N. Janssens, 1–14. Dordrecht: Springer.

Engbers, M. 2018. "Kultur und Differenz in der transdisziplinären Nachhaltigkeitsforschung: Analysen und konzeptionelle Beiträge zur Gestaltung von Lern-/Forschungsprojekten." PhD dissertation, Leuphana University, Lüneburg.

Escobar, A. 2018. *Designs for the Pluriverse: Radical Interdependence, Autonomy, and the Making of Worlds*. Durham, NC: Duke University Press.

Fichant, M. 1975. "Die Epistemologie in Frankreich." In *Geschichte der Philosophie, Band VIII*, edited by F. Châtelet, 118–158. Frankfurt am Main/Berlin/Wien: Ullstein.

Fry, T. 2011. *Design as Politics*. Oxford: Berg Publishers.

Haley, D. 2011. "Ecology and the Art of Sustainable Living." *Field Journal* 4 (1): 17–32.

Harrasser, K., and K. Solhdju. 2016. "Wirksamkeit verpflichtet – Herausforderungen einer Ökologie der Praktiken." *Zeitschrift für Medienwissenschaft* 14 (1): 72–86.

Hörl, E., and O. Leistert, eds. 2019. *Thinking the Problematic: Genealogies, Tracings, and Currents of a Persistent Force*. Bielefeld: transcript.

Jahn, T., M. Bergmann, and F. Keil. 2012. "Transdisciplinarity: Between Mainstreaming and Marginalization." *Ecological Economics* 79: 1–10. doi:10.1016/j.ecolecon.2012.04.017.

Klein, J. T., W. Grossenbacher-Mansuy, R. Häberli, A. Bill, R. W. Scholz, and M. Welti. 2001. *Transdisciplinarity: Joint Problem Solving among Science, Technology, and Society: An Effective Way for Managing Complexity*. Basel: Birkhäuser.

Krohn, W., A. Grunwald, and M. Ukowitz. 2017. "Transdisziplinäre Forschung Revisited: Erkenntnisinteresse, Forschungsgegenstände, Wissensform Und Methodologie." *GAIA* 26 (4): 341–347. doi:10.14512/gaia.26.4.11.

Lang, D., A. Wiek, M. Bergmann, M. Stauffacher, P. Martens, P. Moll, M. Swilling, and C. Thomas. 2012. "Transdisciplinary Research in Sustainability Science: Practice, Principles, and Challenges." *Sustainability Science* 7 (S1): 25–43. doi:10.1007/s11625-011-0149-x.

Maguire, K. 2018. "Transdisciplinarity: Towards an Epistemology of What Matters. ." In *Transdisciplinary Theory, Practice and Education: The Art of Collaborative Research and Collective Learning*, edited by D. Fam, L. Neuhauser, and P. Gibbs, 103–116. Cham: Springer.

Maniglier, P. 2012. "What Is a Problematic?" *Radical Philosophy* 173: 21–23.

Manzini, E. 2015. *Design, When Everybody Designs: An Introduction to Design for Social Innovation*. Cambridge, MA: MIT Press.

Max-Neef, M. 2005. "Foundations of Transdisciplinarity." *Ecological Economics* 53 (1): 5–16. doi:10.1016/j.ecolecon.2005.01.014.

Meyer, E. Forthcoming a. "The Problematic of Transdisciplinary Sustainability Sciences." In Thinking the Problematic: Genealogies, Tracings, and Currents of a Persistent Force, edited by E. Hörl and O. Leistert, Bielefeld: Transcript.

Meyer, E. Forthcoming b. "Solvable Problems or Problematic Solvability? A Collocation Analysis of the Problem Conceptualization in Transdisciplinary Sustainability Sciences and a Possible Contribution from the Philosophy of Science." GAIA 29 (1).

Mitchell, C., D. Cordell, and D. Fam. 2015. "Beginning at the End: The Outcome Spaces Framework to Guide Purposive Transdisciplinary Research." *Futures* 65: 86–96. doi:10.1016/j.futures.2014.10.007.

Nicolescu, B. 2010. "Methodology of Transdisciplinarity – Levels of Reality, Logic of the Included Middle and Complexity." *Transdisciplinary Journal of Engineering & Science* 1 (1): 19–38. doi:10.22545/2010/0009.

Nicolini, D. 2013. *Practice Theory, Work and Organization: An Introduction*. Oxford: Oxford University Press.

Nowotny, H. 2006. *"The Potential of Transdisciplinarity"*. First published in: interdisciplines, http://helganowotny.eu/texts.php

Papanek, V. 1971. *Design for the Real World: Human Ecology and Social Change*. New York: Pantheon Books.

Peukert, D., and U. Vilsmaier. 2019. "Entwurfsbasierte Interventionen in der transdisziplinären Forschung." In *Wege der Vermittlung – Intervention – Partizipation*, edited by M. Ukowitz and H. Renate, 227–250. Wiesbaden: Springer Verlag.

Pohl, C., L. van Kerkhoff, G. H. Hadorn, and G. Bammer. 2008. "Integration." In *Handbook of Transdisciplinary Research*, edited by G. H. Hadorn, H. Hoffmann-Riem, S. Biber-Klemm, W. Grossenbacher-Mansuy, D. Joye, C. Pohl, U. Wiesmann, and E. Zemp, 411–424. Dordrecht: Springer.

Raven, P., and S. Elahi. 2015. "The New Narrative: Applying Narratology to the Shaping of Futures Outputs." *Futures* 74: 49–61. doi:10.1016/j.futures.2015.09.003.

Repko, A. F., and R. Szostak. 2008. *Interdisciplinary Research: Process and Theory*. Thousand Oaks: SAGE.

Savransky, M. 2016. *The Adventure of Relevance: An Ethics of Social Inquiry*. London: Palgrave Macmillan.
Scholz, R. 2011. *Environmental Literacy in Science and Society: From Knowledge to Decisions*. Cambridge, UK: Cambridge University Press.
Schrickel, I. Forthcoming. "The Future of Modern Societies: Problems as Epistemic Design around 1970." In *Thinking the Problematic: Genealogies, Tracings, and Currents of a Persistent Force*, edited by E. Hörl and O. Leistert. Bielefeld: Transcript.
Scott, D. 2014. *Gilbert Simondon's Psychic and Collective Individuation: A Critical Introduction and Guide*. Edinburgh: Edinburgh University Press.
Simon, H. 1969. *The Sciences of the Artificial*. Cambridge, MA: MIT Press.
Simondon, G. 2007. "Das Individuum und seine Genese." In *Struktur, Figur, Kontur: Abstraktion in Kunst und Lebenswissenschaften*, edited by C. Blümle and S. Armin, 29–46. Zürich: diaphanes.
Simondon, G. 2017. *On the Mode of Existence of Technical Objects*. Minneapolis: Univocal.
Stengers, I. 2005. "Introductory Notes on an Ecology of Practices." *Cultural Studies Review* 11 (1): 183–196. doi:10.5130/csr.v11i1.3459.
Stengers, I. 2010. *Cosmopolitics I*. Minneapolis: University of Minnesota Press.
Tejedor, G., and J. Segalas. 2018. "Action Research Workshop for Transdisciplinary Sustainability Science." *Sustainability Science* 13: 493–502. doi:10.1007/s11625-017-0452-2.
Tiles, M. 2012. "What Does Bachelard Mean by Rationalism Appliqué?" *Radical Philosophy* 173: 24–26.
Van Breda, J., and M. Swilling. 2019. "The Guiding Logics and Principles for Designing Emergent Transdisciplinary Research Processes: Learning Experiences and Reflections from a Transdisciplinary Urban Case Study in Enkanini Informal Settlement, South Africa." *Sustainability Science* 14 (3): 823–841. doi:10.1007/s11625-018-0606-x.
Vilsmaier, U. 2018. "Grenzarbeit in integrativer und grenzüberschreitender Forschung." In *Grenzen: Theoretische, konzeptionelle und praxisbezogene Fragestellungen zu Grenzen und deren Überschreitungen*, edited by M. Heintel, R. Musil, and N. Weixlbauer, 113–134. Wiesbaden: Springer VS.
Vilsmaier, U., V. Brandner, and M. Engbers. 2017. "Research In-between: The Constitutive Role of Cultural Differences in Transdisciplinarity." *Transdisciplinary Journal of Engineering & Science* 8: 169–179. doi:10.22545/2017/00093.
Voss, D. 2018. "Simondon on the Notion of the Problem." *Angelaki* 23 (2): 94–112. doi:10.1080/0969725X.2018.1451471.
Welch, J. 2011. "The Emergence of Interdisciplinarity from Epistemological Thought." *Issues in Integrative Studies* 22: 1–39.
Williams, L., T. Bunda, N. Claxton, and I. Mackinnon. 2017. "A Global De-colonial Praxis of Sustainability – Undoing Epistemic Violences between Indigenous Peoples and Those No Longer Indigenous to Place." *The Australian Journal of Indigenous Education* 47 (1): 41–53. doi:10.1017/jie.2017.25.
Wulz, M. 2014. "Technik im Wissen: Zur wechselseitigen Hervorbringung von Wissen, Technik, Geschichte und Gesellschaft in der französischen Wissenschaftsgeschichte und –philosophie." In *Schlüsselwerke der Science & Technology Studies*, edited by D. Lengersdorf and M. Wieser, 54–67. Wiesbaden: Springer VS.

Navigating between Complexity and Control in Transdisciplinary Problem Framing: Meaning Making as an Approach to Reflexive Integration

Basil Bornemann and Marius Christen

ABSTRACT
Referring to a problem-oriented research mode, transdisciplinarity faces the challenge of dealing with the complexity of real-world problems in a methodologically controlled manner. As the first step in transdisciplinary research, problem framing aims at identifying the problem that will serve as a common orientation point for the subsequent research process. Existing approaches to integrative problem framing seek collaboration of practical and scientific actors in order to develop a common definition of the problem. Such a strategy seems to rely on a unifying mode of integration that levels the diversity of perspectives of the involved actors. This paper presents an alternative approach to integrative problem framing in transdisciplinary research, which is based on a mode of reflexive integration. The corresponding approach of 'meaning making' intends to systematically grasp the plurality of perspectives involved and to relate them to an overarching theme. Substantiating and reflecting meaning making based on a research project on sustainability governance in Swiss cantons, the paper argues that meaning making enables a reflexive pathway between complexity and control in transdisciplinary problem framing.

1. Introduction

According to widespread understanding, transdisciplinarity (TD) refers to a specific research mode that brings together scientists and societal actors in joint attempts aimed at solving 'real-world problems' (Klein et al. 2001; Zierhofer and Burger 2007). The approach has been especially heralded as adequate to generate knowledge for tackling complex and 'wicked' sustainability problems in a methodologically controlled way (Lang et al. 2012; Brandt et al. 2013). Within an ideal-typical multistage process of transdisciplinary research, problem framing is the first step. Problem framing is concerned with identifying and delineating those problems that ought to be addressed and potentially solved by a research project (Pohl and Hirsch Hadorn 2007; Lang et al. 2012; Bergmann et al. 2012; Polk 2015). Therefore, problem framing is a particularly decisive and path-defining phase in TD research: by clarifying what the project is all about (or not), problem framing has the potential to shape the subsequent research process in terms of its objects, goals, and methods as well as the actors involved (Rossini 2009).

Due to its formative function in the research process as a whole, problem framing in particular faces the challenge of following two general orientations of TD. In order to do justice to transdisciplinary *problem orientation*, problem framing must on the one hand adequately reflect the complexity of the real-world problem; in order to satisfy the transdisciplinary *research orientation* and to produce a researchable problem, it must on the other hand be carried out in a methodologically controlled manner. A deviation to one side or the other is problematic. Excessive emphasis on complexity and an ideally complete representation of the real-world problem harbor the risk of overwhelming research or even abandoning the scientific character of a TD project altogether. Excessive scientific control, on the other hand, potentially comes at the expense of the representation of complexity – and of openness to discovery and contextuality; it potentially implies the danger of blocking the production of problem-specific knowledge and corresponding innovative solutions. Therefore, TD problem framing must navigate between the Scylla of paralyzing complexity and the Charybdis of excessive control – but how so?

In this paper, we address this question by presenting and reflecting a specific approach to TD problem framing, which we call 'TD meaning making.' Proceeding from the assumption that navigating between complexity and control requires an integrative approach, TD meaning making is introduced as a specific proposal on how this integration might take place. While established approaches to TD problem framing seem to be based on a unifying mode of integration that levels differences in perspectives, TD meaning making employs a reflexive mode of integration that maintains diversity in unity. More specifically, meaning making is supposed to reveal the diversity of scientific and practical understandings of a complex and contested subject matter and put the different understandings into productive relations. It thereby opens the view for common but differentiated problems instead of unifying perspectives toward a shared definition of the problem. We developed and applied this approach of meaning making in a TD project addressing sustainability governance in Swiss cantons, which we carried out as an interdisciplinary research team in collaboration with sustainability-oriented practitioners of cantonal administrations.

We first briefly describe the theme – that is, sustainability governance – as well as the background and goals of the research project and explain why we opted for a TD approach (section 2). Next, we characterize TD as problem-oriented research facing the challenge of navigating between complexity and control and examine a prominent approach to TD problem framing that promotes integration as a means of dealing with the challenge (section 3). Based on concerns about this proposition, we then present meaning making as an alternative, reflexive approach to integrative problem framing and illustrate how we applied it in our TD project on sustainability governance (section 4). After reflecting on the implications and potentials of meaning making (section 5), the paper concludes with some propositions on how to further develop the theory and methodology of TD meaning making in particular and TD problem framing in general (section 6).

2. A Transdisciplinary Project on Sustainability Governance

Sustainability governance (SG) generally can be understood as the organization of collective action toward sustainability (Meadowcroft 2007). Specifically, it involves the design and implementation of governing practices – from central hierarchical control to the promotion of stakeholder participation – that enable sustainability-oriented problem solving and social change. In this way, SG is considered a highly ambitious endeavor that fundamentally challenges established programmatic, institutional, and procedural conditions of governance. For example, SG is supposed to overcome the short-termism that dominates policy making toward a more long-term perspective, or to improve the integration of policies in order to tackle the complexity of sustainability problems (Newig, Voss, and Monstadt 2008; Bornemann 2014; Steurer 2010).

Sustainability governance is associated with various kinds of complexity. Conceptually, SG is complex in that it combines two complicated concepts – governance and sustainability – and therefore includes descriptive assumptions about the actual functioning of governance systems as well as normative assumptions about their intended (sustainability-oriented) functioning (Lange et al. 2013). From a practical point of view, SG is complex in that SG arrangements consist of numerous actors (with different ideas and interests) as well as institutional (polity), substantial (policy), and procedural (politics) elements that are highly interdependent (Bornemann and Christen 2019). Overall, SG refers to a hybrid field that combines conceptual elements of systematic 'knowing' with practical elements and experience of contextual 'doing' (Bornemann, Lange, and Burger 2019).

Despite initial successes of governments and public administrations to integrate sustainability in their governance practice (Lafferty and Meadowcroft 2000; Jacob and Volkery 2004), practitioners struggle with promoting sustainability-oriented governance transformations 'on the ground'. More specifically, practitioners seem to encounter problems when it comes to making SG an important orientation for government practice as a whole. On the other hand, although preconditions and conditions as well as approaches and strategies of SG are discussed extensively in science, existing analyses often seem to inadequately address the problems existing in practice. Scholars tend to focus on general principles, formal institutions, strategies, and instruments, but have little knowledge of how practitioners perceive and perform SG 'on the ground', under what conditions they work, and what daily challenges they face (Fiorino 2010). In contrast, public administrators see themselves confronted with abstract theoretical SG principles or mechanical 'toolbox advice,' both of which are detached from their working contexts.

Driven by the motivation to identify problems related to the design and practice of SG in public administration, we conducted a research project on SG in Swiss cantons in collaboration with sustainability practitioners from several cantonal administrations. The project was carried out by an interdisciplinary research team with researchers from the fields of sustainability and political science as well as philosophy. The sustainability practitioners included the heads of six cantonal sustainability offices, who bear a central responsibility for sustainability in the public administration of their respective cantons (Aargau, Basel-Stadt, Bern, St.Gallen, Vaud, and Valais; for more details see Bornemann and Christen 2019). The project's overall goal was not to generate solutions, but to establish an agenda of problems that would be further addressed in a follow-up research project. Although the impetus for the project came from scientists, its basic motivation to improve SG with regard to relevant problems was shared by the involved practitioners.

To identify problems related to SG in cantonal administrations, we considered it crucial to establish a close exchange between practice and science. Accordingly, we did not merely want to treat SG as an object of scientific research, to be analyzed and, eventually, improved by scholarly insights. Rather, we regarded both science and practice as relevant sources of different kinds of knowledge that can mutually illuminate each other when it comes to the identification of problems related to SG. Therefore, we deemed TD a promising approach to systematically relate practical and scientific perspectives on equal footing (Scholz and Steiner 2015; Polk 2015).

Before we introduce and discuss the specific approach to TD problem framing that we have applied in our project, the following section reflects the challenges associated with TD problem framing on a theoretical level.

3. Transdisciplinary Problem Framing between Complexity and Control

Transdisciplinarity research is regularly qualified, if not identified, as *problem-oriented research*, in contrast to, for example, curiosity-driven basic research (Bergmann et al. 2012; Schmidt 2011). Accordingly, TD combines two ambitions: the orientation toward real-world problems on the one hand and the idea of some form of methodologically systematic research on the other.

The centrality of *problems* in TD research is reflected in the fact that real-world problems are seen as the 'starting point for research' (Bergmann et al. 2012, 27). Real-world problems (or 'life-world problems') are usually referred to as problematic properties of social reality experienced by actors prior to theorization (Pohl and Hirsch Hadorn 2008, 428). They are 'characterized by complex structures and uncertain knowledge bases' (Bergmann et al. 2012, 31). In a systematic account, Schmidt (2011) refers back to a classical analytical definition, according to which a problem denotes a difference between an actually existing and a desired state of the world. Following this understanding, we conceive a problem as a name given to a constellation of issues, where a given state of affairs deviates from a desired state. This implies that there is actually not such a thing as an objective problem out there in the real world. Denoting a gap between a given state (*what is the case*) and a desired state (*what shall be the case*), problems must be conceived of as potentially disputed social constructions consisting of descriptive 'system knowledge' and normative 'target knowledge.'

Regardless of some objectivist flickering (*real-world* problems), this social constructivist notion of problems is also supported by the term problem *framing* (instead of problem *definition*), which is mostly used in TD literature and emphasizes the *social* qualities of a problem-identification *process* (Bergmann et al. 2012, 36; Polk 2015, 112). Problem framing, consequently, is shaped by the social settings in which it takes place, and problems most probably will be articulated differently by various societal groups, given their diverse involvement in and perceptions of the given situation as well as their different norms and values. Accordingly, in TD, the real-world problem is understood as a social construction on the basis of which knowledge production is organized. As such, it is also a means of social cohesion, a thing around which actors can gather and that can serve as a commonly accepted reference point for a potentially conflictual process of problem solving ahead (Bergmann et al. 2012).

The self-description as problem-oriented *research*, however, expresses that TD is not just any kind of problem-oriented activity. Rather, the reference to research points out that it is activity oriented and organized according to scientific standards and norms, differing from other problem-oriented approaches in management or politics. Specifically, TD research is a way of addressing the real world in a 'scientifically controlled and reflective manner' (Bergmann et al. 2012, 27) – even if the participation of nonscientific actors in the research process as another constitutive element of TD entails a specific mode of research (Zierhofer and Burger 2007; Hirsch Hadorn et al. 2008a; Jahn, Bergmann, and Keil 2012). The scientific aspiration is expressed in the fact that TD is oriented toward (more or less) standardized and generally accepted procedures and methods, from disciplinary methods to models for structuring the overall process and concrete techniques of collaborative knowledge production (Defila and Di Giulio 2015). Therefore, TD is referred to as a principle of *science* that reacts to changing epistemic demands due to problem orientation on the one hand and involvement of nonscientific actors on the other (Mittelstraß 2005).

In view of this understanding of TD as problem-oriented research, TD problem framing as the first and formative step in the research process faces the challenge of mediating between two opposing poles: on the one hand, it needs to do justice to the complexity of real-world problems and satisfy the demands of involved actors for problem-specific and contextualized solutions; on the other hand, problem framing needs to do justice to the need of a systematic and methodologically controlled research process and, in particular, to generating a well-constituted object that can be researched. Thus, TD problem framing can be interpreted as a tension-filled process navigating between doing justice to real-world complexity and scientific control.

Examining the existing TD literature on how this navigation can be conceived and practiced in the phase of problem framing, we find, in addition to more case-related approaches (see Hirsch Hadorn et al. 2008b; Scholz, Spoerri, and Lang 2008; Schäfer and Kröger 2016), some generalizing methodological considerations (Lang et al. 2012; Polk 2015). We take a closer look at a exemplary contribution that makes a concrete proposal for navigating between the two poles. In their widespread methodological and practical handbook, Bergmann et al. (2012, 31–36) point to three approaches to problem framing: The 'science-focused approach' implies that the problem is identified based on objective scientific criteria and analysis of the situation. In contrast, the 'real-world-focused approach' implies that the formulation of the problem can be delegated to the social actors involved in a TD project. As a middle way, the authors argue for a third, 'integrative approach'. Here, both scientists and practitioners have a say in co-creating the problem. In view of the dual ambition of problem-oriented research and the associated challenge of navigating between complexity and control, an integrative approach seems to provide an appropriate basis for TD problem framing.

Bergmann et al. (2012, 37–38) describe integrative problem framing as an iterative process of intensive discussions among the involved actors. Starting from a 'first, provisional description of the societal problem,' the scientific disciplines 'appropriate to the task' and 'the necessary practical knowledge' are supposed to be 'analyzed discursively.' In a process characterized by 'intensive discussion' within the 'emerging team,' the composition of the involved actors as well as the 'key aspects of the societal problem' are developed and constantly reviewed and revised. In addition to identifying the key aspects of the problem at hand, this step serves the exchange of 'alternative views of the problem' and the working out of differences among the involved actors. This results in a common agreement, by both scientists and practitioners, 'on an exact description of the problem.' Based on this 'common view of the problem,' it may turn out to be necessary 'to review the makeup of the team [...] with respect to both the scientific fields and societal actors involved,' implying a new common description and working out of the problem.

Overall, the integrative approach to TD problem framing is expected to lead to a precisely and comprehensively defined problem that all parties from science and practice agree on. With regard to a general understanding of integration as the connecting of individual elements (integrands) to an integral whole in different ways (Bornemann 2014), Bergmann et al. seem to base their approach on a specific mode of integration that can be described as 'unifying integration.' Its aim is to solve or even harmonize the differences between the integrands (i.e., different scientific and social problem articulations) and to form an integral (i.e., a common problem description) in which the differences have been resolved in favor of one clear problem. Such a unifying mode of integration, expressed above all in the declared goal of developing a common view of a single problem, comes with both epistemic and social consequences for TD research.

In epistemic terms, by resolving or leveling differences between problem views, a unifying integration mode comes with the risk of narrowing innovative solution approaches. Practitioners from diverse contexts, but also scientists from different disciplines, enter a process of problem solving with specific background knowledge and beliefs. On the basis of these different horizons of interpretation (Hirsh 2013) they develop individual perceptions and insights about what is problematic. When TD problem framing aims at a definition of a common problem – instead of a clarification of the problem perspectives – variations or even contradictions in problem perceptions might move into the background. This can lead to relevant aspects of complex real-world problems not being reflected upon or only insufficiently considered. Moreover, the notion of a common problem definition implies a fundamental compatibility or even complementarity between different aspects of a problem (or their perception) that ultimately form a coherent whole; that is, a common problem. This neglects fundamental normative or cognitive conflicts that may not be unveiled and integrated into a shared problem perspective. Finally, the orientation of an entire TD project toward a common problem may not only block open dynamics in context-oriented and innovative knowledge production but also entail a solutionist logic (Strohschneider 2014).

In social terms, a unifying mode of integration is based on the premise that actors, both practitioners and academics, are actually willing and able to agree on a common problem definition. This may be overburdening, as the actors potentially have ideas and interests that cannot be easily unified. Attempts to align the actors to a common understanding of the problem may disregard their specific perceptions and, thus, potentially impair the legitimacy of the entire TD process. Moreover, if those with the strongest social positions and the greatest resources (time, money, rhetorical skills, etc.) capture the problem framing to put their perspectives in the foreground, this could lead to a hegemonic problem definition – that is, to a definition of what is and should be, based on the perceptions and preferences of only a few actors involved.

In light of these concerns related to epistemic and social aspects of a unifying mode of integration, we argue that it might be fruitful not to immediately immerse in the identification and definition of a common problem by consolidating the diversity of disciplinary and stakeholder perspectives. Hence, in the following section we present an alternative approach to integrative TD problem framing.

4. Meaning Making as a Reflexive Approach to Transdisciplinary Problem Framing

Instead of considering integration as unification, our approach is based on a reflexive mode of integration that aims to capture the complexity of real-world problems in a methodologically controlled way, so that the diversity of problem perspectives is brought together and linked in an integral view without dissolving the differences. Based on some general thoughts on reflexive integration, we outline our approach of TD meaning making (4.1). We then explain how we applied this approach in our project on cantonal SG (4.2).

4.1. Transdisciplinary Meaning Making: The General Approach

In view of the above outlined conceptual understanding of problems as deviations of what is and what should be, problem perceptions do in fact vary a lot. Constructing and articulating problems on the basis of different interpretation horizons particularly applies to problems that are located in the social world, such as problems related to the design and implementation of governance, as they are fundamentally based on social constructions and interpretations (Yanow 2009). Assuming that diversity provides access to real-world complexity, we advocate a form of integration that embraces this diversity of problem articulations. To enhance transparency and conceptual clarity, the proposed kind of reflexive integration begins by revealing and making explicit the different perspectives and assumptions that are constitutive of the perceived problem(s). Unfolding the space of possible problem perceptions allows for identifying the varieties of problems associated with a specific theme. By establishing a balance between differentiated integrands and the integral whole, it aims at keeping differences alive under a common point of reference, such as, in our case, the notion of SG.

We operationalize this form of integration in an approach we call 'meaning making.' Informed by interpretive social science methodology (Yanow 2009; Yanow and Schwartz-Shea 2006; Wagenaar 2011), meaning making highlights conceptual clarification as a pivotal and initial step in TD problem framing. It serves to reveal the interpretation horizons against which problems are articulated – that is, the experiences, knowledge, value perspectives, and beliefs on which actors interpret what is and what shall be the case in relation to a particular field or topic. Making explicit these diverse descriptive and normative assumptions constituting the problems allows for revealing the subject matter's complexity. Differentiating descriptive system and normative target knowledge might be understood as a strategy to reduce complexity (see Pohl and Hirsch Hadorn 2007, 36). However, it could also be understood as a way to unfold the complexity of a topic. Committed to a logic of 'opening' rather than 'closing' or 'narrowing' (Stirling 2008), the approach of meaning making, as we propose it, does not aim to produce a clear definition of a problem at hand. Rather, it is about capturing different and also opposing – practical as well as scientific – views and maintaining them in the ongoing research process. Hence, TD meaning making does not follow the claim that consensus among all actors involved must be reached but that differences should be made explicit (similarly Ukowitz 2014; Pohl 2014).

The gaps, overlaps, and tensions that become visible through this approach present a 'universe of meanings' which spans the space of possible and existing understandings of a topic and in which differences do not have to be leveled or adjusted. They rather clarify the challenges in the subject field and thus prepare the ground for the identification of problems. In the sense of a reflexive approach, meaning making is therefore, on the one hand, about revealing and reflecting the assumptions and understandings upon which actors construct problems. On the other hand, the universe of meanings itself becomes a reflexive point of reference for the collective identification and construction of problems. In addition to these epistemic functions, making assumptions transparent also serves as a social means for controlling and overcoming actors' partial interests. This works not only as an approach to prevent the research agenda from being one sided but also to strengthen the social cohesion in a TD project.

In practice, meaning making begins with collecting and explicating the descriptive and normative assumptions of the actors involved on a certain topic or concept that serves as a common point of reference. In a structured and moderated dialogue, these articulations of individual understandings, which guide the problem perceptions, are explicated, compared, and related to each other. Thereby, the universe of meanings is constructed in form of an encompassing framework of possible understandings of the respective topic. This framework allows differences to be discussed on a transparent, well-informed, and comprehensive basis. Tensions and contradictions between the descriptive and normative assumptions serve to identify generic and context-specific problems.

Overall, meaning making aims for capturing diversity in unity in a structured process. In the next section, we illustrate how we operationalized and applied meaning making in our TD project about SG in Swiss cantons.

4.2. TD Meaning Making: An Exemplary Application

In order to identify problems related to the design and implementation of SG in the context of Swiss cantonal administrations, we have applied our general approach of TD meaning making in the form of five concrete steps.

4.2.1. Setting the Scene: Theoretical Structuring

We started the inquiry from a scientific point of view by developing basic theoretical reference points for conceiving and grasping practical understandings of SG on the ground. Reflecting the state of the scientific discussion, we derived two types of theoretical references from a review of the relevant theoretical, empirical, and practical literature. First, to grasp the normativity in the field (i.e., references to what SG should accomplish), we compiled an overview of normative principles and requirements in relation to SG. Second, in order to examine the current situation of SG on the ground (i.e., the actual state: how SG is implemented), we developed a descriptive analytical framework that focuses on sustainability-oriented practical governance arrangements (Bornemann and Christen 2019). We have kept this analytical framework open and abstract in order to capture the diversity of the cantonal sustainability officers' SG understandings and apprehend the respective implicit normative and descriptive assumptions as authentic as possible.

4.2.2. Grasping Diversity: Document Study and In-depth Interviews

The second step was to gather a rich information base about SG in the selected cantonal administrations. Drawing on cantonal communication materials and policy documents, we first obtained an overview of the basic self-understandings, organizational structures, and activities in relation to SG. Next, we conducted semi-structured qualitative interviews with cantonal sustainability officers to collect their understandings, actual practices, and perceived challenges regarding SG. Due to their official mandates, they have rich, context-specific knowledge of conditions, forms, and meanings of SG in public administrations. In addition to open questions to capture the diverse practitioners' perceptions of real-world issues in an open manner, the interview guide also included focused questions based on the theoretical reference points.

4.2.3. Portraying Complexity: Analysis of SG Understandings

In a third step, we organized this rich material with the help of a qualitative-interpretive approach. First, we used a code-based content analysis (Krippendorff 2013) to identify and encode those text elements that correspond to the theoretical reference points developed in step one. Beginning with categories that were initially very general, we further refined them in the course of the analysis. Based on the fully coded text material, we compiled a comprehensive list of all coded elements for each canton. Next, we cautiously condensed the individual characteristics on each list through intensive discussion and summarized the central elements in the form of canton-specific narrative profiles on understandings, practices, and challenges related to SG. Both the list of individual characteristics and the narrative profiles were compiled into working papers that formed the basis for two transdisciplinary workshops. The collection of these portraits represents the totality of SG meanings in the cantons studied, covering both descriptive and normative aspects.

4.2.4. Debating Complexity: Transdisciplinary Workshops

At the heart of our process of TD problem framing were two half-day workshops in which both cantonal sustainability officers and scientists took part. In addition to facilitating mutual exchange between the participants, the workshops aimed to further refine the scientific and practical meanings of SG, to display them in a 'universe of meanings of SG' in the cantonal context of Switzerland, and to identify relevant challenges on this basis.

The first workshop focused on the descriptive and normative assumptions underlying the different views of SG. First, the cantonal portraits were presented and each participant could ask clarifying questions. Then, a member of the research group put forward for discussion an interpretation of the portraits, revealing similarities and differences between them with regard to the scientific debate. Subsequently, the normative and descriptive assumptions about SG in the various cantons were related to each other in an open discussion. The resulting universe of meanings provided a comprehensive and differentiated overview of congruent and contradictory interpretations of SG, beyond a common definition.

Based on this, the second workshop aimed to identify challenges related to the design and implementation of SG in the cantons. After first discussing the SG-related challenges and potentials articulated by the practitioners on the basis of the cantonal portraits, a member of the research team again presented an interpretation of the practical perspectives on challenges with reference to the scientific discussion on SG. On the one hand, this analysis revealed overlaps between scientific and practical challenges, such as the discrepancy between the goal of integrative policy-making and the prevailing compartmentalization of the political decision-making system. On the other hand, questions were raised that are hardly discussed in scientific literature so far. For example, practitioners stressed the incompatibility of the complex sustainability idea with communication culture in public administration and underlined the importance of 'issue framing' as a relevant activity in their attempts to 'sustainabilize' administrative practice. In a third part, all participants – practitioners and scientists – identified connections in the form of overlaps and tensions between the challenges with regard to the universe of SG meanings. In this way, general challenges could be filtered out, such as issues of accountability or sustainability as a difficulty in bulky contexts. Finally, in a moderated assessment, the participants identified the most relevant challenges – that is, those challenges that are considered particularly urgent from a practical point of view and that can be adequately articulated within a scientific SG framework.

4.2.5. Framing the Problem: Scientific Generalization

The last step aimed to bring together the concrete challenges related to SG in an overarching problem framework. In view of the purpose of the project, which was to create the basis for a scientific follow-up project, this problem framework should, on the one hand, be connected to existing scientific debates. On the other hand, it should cover and relate the numerous individual practical problems articulated by practitioners in relation to the design and implementation of SG in cantonal administrations. In order to meet both requirements, we proposed, from a scientific point of view, to place the identified challenges of cantonal SG in the context of the more recent scientific discussion on the sustainable state. Accordingly, we proposed an overarching (generic) problem framework to the practitioners, to which all problem views and challenges of SG identified in the project could refer. Accordingly, SG was framed as (self-)transformation of actually existing state governance toward a normative model of the sustainable state. This problem was condensed into a generic research question: How are states transforming themselves toward sustainability? This overarching framework of problems was translated into a series of empirical research questions, which in turn were linked to the specific practical challenges identified in the project. To ensure that the understandings and challenges regarding SG were adequately taken into account, the problem framework and the specific research questions were discussed and approved in a concluding working session with practitioners.

5. Reflections on TD Meaning Making as an Approach for Reflexive Integration

By using meaning making in our TD problem framing project on cantonal SG, we were able to identify an overall problem framework that was considered relevant by both the involved scientific and practical actors, precisely because it captures and links their different context-specific understandings and challenges. In view of this case-specific success, we consider meaning making a promising approach to integrative TD problem framing more generally. It can be seen as an approach to reflexive integration in so far as it navigates between the two poles of TD research – complexity of real-world problems and scientific control – not by unifying differences but maintaining them within a generic whole.

Overall, meaning making made it possible to grasp the complexity of the subject by comprehensively uncovering and collating the manifold views and assumptions regarding SG. Instead of narrowing the perspective from the outset to a shared problem by pushing the actors into a mode of problem formulation, meaning making promotes opening the perspectives and a kind of 'thinking outward' (Huutoniemi and Tapio 2014). It primarily aims at explicating and clarifying normative and factual assumptions and expectations on the basis of which problems are perceived. The activation and disclosure of different views on SG and their integration into a universe of meaning proved to be helpful in enabling an open discussion among the participating practitioners and scientists. On this basis, it was possible to overcome existing problem perspectives and to adopt innovative ones. For example, when it became clear that cantons have very different context-specific SG practices, actors expressed concerns about the dominant approach to compare cantonal practices. Instead of just learning from each other, cantonal practices should also be interpreted and evaluated with regard to overarching but contextually interpretable sustainability goals, that is, for example, with regard to a conceptually based understanding of the sustainable state.

In order to enable this recognition of complexity, meaning making also relies on moments of systematization and control. For example, the theoretical reference points developed in step 1 made it possible to comprehensively collect the perceptions of the individual actors, to gain meaningful narratives about the various interpretations and practices of SG in the cantons, and to associate these different practical findings with scientific meanings. Although we have kept the scheme open in order to capture the actors' understanding as unfiltered as possible, it figured as a means of epistemic control based on preconceived ideas for structuring a theme – and made a constructive and controversial dialogue about SG possible in the first place. Additionally, the methodological design includes a number of control instances. They range from the selection of cantons and interview partners through the design of interview guidelines and interview situations to the creation of categories for analyzing the material. In addition to these methodical control tools, a thorough organization and a goal-oriented, albeit essentially open, moderation of the process – especially the workshops (step 4) – proved to be an important factor in steering expectations and communication dynamics.

This points to some additional insights about the social implications and consequences of meaning making. In addition to its potential for reflexive integration in epistemic terms, meaning making also contributes to social integration. The explicit consideration of individual perceptions and experiences of all actors involved alleviated the problems of (perceived) asymmetry and mutual skepticism between scientists and practitioners that are a typical challenge for collaborative projects. For example, while the interview phase (step 2) was essentially devoted to generating a comprehensive overview of the practical understandings of SG, it also proved to be an important element in building a safe space for open articulation of individual perspectives. The relative openness of the interpretative approach (step 3) also had a trust-building effect, as it ensured that the views and experiences of practitioners were taken seriously and accepted by scientists.

Besides these promising epistemic and social aspects of meaning making, we also see some indications for challenges, for example, with regard to the roles and expectations of the actors involved. While we saw our role as scientists in creating a space for debate aimed at opening and recognizing real-world complexity, we found that practitioners repeatedly expected us scientists to reduce complexity by defining concepts and problems. Conversely, we as scientists were caught up in the expectation that practitioners were interested only in pragmatic and action-oriented, rather than systematic knowledge. This assumption turned out to be misleading, for example, when practitioners questioned the theoretical assumptions of the SG's scientific understanding or when they stressed the need to reflect on normative questions relating to SG. Although we have consciously paid attention to the supposed roles of the different actors in the process design, we have learned that the specific roles of the actors involved in problem framing are more diverse and complex than expected.

Overall, we believe that meaning making is a promising way to identify new problems related to complex and unclear issues in a socially inclusive and robust way. By embracing complexity in a controlled way – that is, by employing a mode of reflexive integration – it has the potential to create contextual frameworks for real-world problems that are also scientifically researchable. In our case, the ongoing research on the challenge of the sustainable state has yet to show whether and how the resulting problem framework offers promising starting points for innovative solutions.

6. Conclusion and Outlook

In this paper we have argued that problem framing, as the first and forming phase of a TD project, is particularly faced with the challenge of navigating between two poles that orient TD in general. On the one hand, it has to take account of the complexity of real-world problems in order to enable problem-specific and context-adequate solutions. On the other hand, it must meet the need for scientific control in terms of a systematic process of knowledge production. Existing approaches of TD problem framing seem to resolve this tension by embarking on a unifying mode of integration that seeks a common problem description. We believe that this comes at the cost of losing the potential to generate innovative problem perspectives through the confrontation and relation of different contextual perceptions of a subject matter. As an alternative approach to TD problem framing, we presented TD meaning making, which relies on a reflexive logic of integration that preserves diversity in unity. We have operationalized TD meaning making and applied it to a project on SG in Swiss cantons. On this basis, we have shown how meaning making is a scientifically controlled approach that takes into account the different perspectives and experiences of practitioners, as well as the different scientific interpretations, and integrates them in a universe of meanings that serves as a basis for identifying various problems. In our case, TD meaning making proved to be a fruitful approach to navigate between the Scylla of complexity and the Charybdis of control.

In view of our findings and experience, we see three main routes to further develop the theory and methodology of meaning making in particular as well as the debate about TD problem framing in general. First, we believe that, in addition to further practical tests, the approach of meaning making is also worth more thorough reflections with regard to its methodological underpinnings. As mentioned above, we drew on interpretive methods, but we still see the need to further elaborate the approach by discussing scientific and practical variants and their implications and by linking it to specific methods such as frame analysis or narrative analysis.

With respect to the significance of the social dimension of meaning making, a second promising path for future work is to develop a more differentiated (conceptual and empirical) understanding of expectations and roles of both practitioners and scientists (Pohl et al. 2010; Wittmayer and Schäpke 2014). Such clarification could be of practical value as it could, for example, serve to keep roles in TD processes transparent and to explicate envisioned (self-)expectations.

We are well aware that there are many different topic fields with different characteristics as well as other forms of thinking and practicing integration within TD problem framing. This, finally, raises some interesting questions: Under what circumstances should we choose which approach to TD problem framing? When should we opt for more control, and when should we allow complexity to propel the research process? Which approach is best suited to which aims and conditions and with what effects? We believe that there is no 'one-size-fits-all' solution. Rather, the adequate mode of integration – be it unifying, reflexive, or another mode – must be chosen in relation to specific fields, actors, aims, and contexts.

Acknowledgments

We are indebted to Marlène Schenk for effective support in carrying out the project. Further, we extend our thanks to our project partners from the mentioned Swiss cantons for taking part in the process as well as to the editors of the special issue and two reviewers for their helpful comments.

Disclosure statement

No potential conflict of interest was reported by the authors.

Funding

This work was supported by the 'Sustainable Development at Universities Programme' from (former) Swiss University Conference (SUC) under Grant C1-38.

ORCID

Basil Bornemann http://orcid.org/0000-0002-9435-1500

References

Bergmann, M., T. Jahn, T. Knobloch, W. Krohn, C. Pohl, and E. Schramm. 2012. *Methods for Transdisciplinary Research: A Primer for Practice*. Frankfurt A.M.: Campus.

Bornemann, B. 2014. *Policy-Integration und Nachhaltigkeit: Integrative Politik in der Nachhaltigkeitsstrategie der deutschen Bundesregierung*. 2nd ed. Wiesbaden: Springer VS.

Bornemann, B., and M. Christen. 2019. "Sustainability Governance in Public Administration: Interpreting Practical Governance Arrangements in Swiss Cantons." *Environmental Policy and Governance* 29: 159–169. doi:10.1002/eet.v29.3.

Bornemann, B., P. Lange, and P. Burger. 2019. "Toward Systematic Understandings of Sustainability Governance. A Conceptual Meta-Framework." In *Sustainability Governance and Hierarchy*, edited by P. Hamman, 60–78. London: Routledge.

Brandt, P., A. Ernst, F. Gralla, C. Luederitz, D. J. Lang, J. Newig, F. Reinert, D. J. Abson, and H. von Wehrden. 2013. "A Review of Transdisciplinary Research in Sustainability Science." *Ecological Economics* 92: 1–15. doi:10.1016/j.ecolecon.2013.04.008.

Defila, R., and A. Di Giulio. 2015. "Integrating Knowledge: Challenges Raised by the 'Inventory of Synthesis'." *Futures* 65: 123–135. doi:10.1016/j.futures.2014.10.013.

Fiorino, D. J. 2010. "Sustainability as a Conceptual Focus for Public Administration." *Public Administration Review* 70: 78–88. doi:10.1111/j.1540-6210.2010.02249.x.

Hirsch Hadorn, G., S. Biber-Klemm, W. Grossenbacher-Mansuy, H. Hoffmann-Riem, D. Joye, C. Pohl, U. Wiesmann, and E. Zemp. 2008a. "The Emergence of Transdisciplinarity as a Form of Research." In *Handbook of Transdisciplinary Research*, edited by G. H. Hadorn, H. Hoffmann-Riem, S. Biber-Klemm, W. Grossenbacher-Mansuy, D. Joye, C. Pohl, U. Wiesmann, and E. Zemp, 19–39. Dordrecht: Springer.

Hirsch Hadorn, G., H. Hoffmann-Riem, S. Biber-Klemm, W. Grossenbacher-Mansuy, D. Joye, C. Pohl, U. Wiesmann, and E. Zemp. 2008b. *Handbook of Transdisciplinary Research*. Dordrecht: Springer.

Hirsh, J. B. 2013. "Meaning and the Horizon of Interpretation: How Goals Structure Our Experience of the World." In *The Experience of Meaning in Life*, edited by J. A. Hicks and C. Routledge, 129–139. Dordrecht: Springer.

Huutoniemi, K., and P. Tapio. 2014. *Transdiciplinary Sustainability Studies. A Heuristic Approach*. New York: Routledge.

Jacob, K., and A. Volkery. 2004. "Institutions and Instruments for Government Self-Regulation: Environmental Policy Integration in a Cross-Country Perspective." *Journal of Comparative Policy Analysis* 6 (No. 3): 291–309. doi:10.1080/1387698042000305211.

Jahn, T., M. Bergmann, and F. Keil. 2012. "Transdisciplinarity: Between Mainstreaming and Marginalization." *Ecological Economics* 79: 1–10. doi:10.1016/j.ecolecon.2012.04.017.

Klein, J. T., W. Grossenbacher-Mansuy, R. Häberli, A. Bill, R. W. Scholz, and M. M. Welti, eds.. 2001. *Transdisciplinarity: Joint Problem Solving among Science, Technology, and Society: An Effective Way for Managing Complexity*. Basel: Birkhauser.

Krippendorff, K. H. 2013. *Content Analysis – 3rd Edition: An Introduction to Its Methodology*. Thousand Oaks: Sage.

Lafferty, W. M., and J. Meadowcroft. 2000. "Patterns of Governmental Engagement." In *Implementing Sustainable Development. Strategies and Initiatives in High Consumption Societies*, edited by W. M. Lafferty and J. Meadowcroft, 337–421. Oxford: Oxford University Press.

Lang, D., A. Wiek, M. Bergmann, M. Stauffacher, P. Martens, P. Moll, M. Swilling, and C. J. Thomas. 2012. "Transdisciplinary Research in Sustainability Science: Practice, Principles, and Challenges." *Sustainability Science* 7 (1): 25–43. doi:10.1007/s11625-011-0149-x.

Lange, P., P. J. Peter, A. S. Driessen, B. Bornemann, and P. Burger. 2013. "Governing Towards Sustainability—Conceptualizing Modes of Governance." *Journal of Environmental Policy & Planning* 15 (3): 403–425. doi:10.1080/1523908X.2013.769414.

Meadowcroft, J. 2007. "Who Is in Charge Here? Governance for Sustainable Development in a Complex World." *Journal of Environmental Policy & Planning* 9 (3): 299–314. doi:10.1080/15239080701631544.

Mittelstraß, J. 2005. "Methodische Transdisziplinarität." *Technikfolgenabschätzung – Theorie und Praxis* 2 (14): 18–23.

Newig, J., J.-P. Voss, and J. Monstadt, eds. 2008. *Governance for Sustainable Development: Coping with Ambivalence, Uncertainty and Distributed Power*. London: Routledge.

Pohl, C. 2014. "Eine Theorie Transdisziplinärer Forschung für wen? Reaktion auf M. Ukowitz. 2014. Auf dem Weg zu einer Theorie transdisziplinärer Forschung GAIA 23/1: 19–22." *GAIA – Ecological Perspectives for Science and Society* 23 (3): 216–220. doi:10.14512/gaia.23.3.3.

Pohl, C., and G. Hirsch Hadorn. 2007. *Principles for Designing Transdisciplinary Research*. München: oekom.

Pohl, C., and G. Hirsch Hadorn. 2008. "Core Terms in Transdisciplinary Research." In *Handbook of Transdisciplinary Research*, edited by G. H. Hadorn, H. Hoffmann-Riem, S. Biber-Klemm, W. Grossenbacher-Mansuy, D. Joye, C. Pohl, U. Wiesmann, and E. Zemp, 427–432. Dordrecht: Springer.

Pohl, C., S. Rist, A. Zimmermann, P. Fry, G. S. Gurung, F. Schneider, C. I. Speranza et al. 2010. "Researchers' Roles in Knowledge Co-production: Experience from Sustainability Research in Kenya, Switzerland, Bolivia and Nepal." *Science and Public Policy* 37 (4): 267–281. DOI:10.3152/030234210X496628.

Polk, M. 2015. "Transdisciplinary Co-production: Designing and Testing a Transdisciplinary Research Framework for Societal Problem Solving." *Futures* 65: 110–122. doi:10.1016/j.futures.2014.11.001.

Rossini, M. 2009. "Was Ist Das Problem? Problemstrukturierung in der inter- und transdisziplinären Forschung. Bericht von der 'Transdisciplinary Conference 2018'." *Technikfolgenabschätzung – Theorie und Praxis* 1 (18): 117–119.

Schäfer, M., and M. Kröger. 2016. "Joint Problem Framing in Sustainable Land Use Research. Experience with Constellation Analysis as a Method for Inter- and Transdisciplinary Knowledge Integration." *Land Use Policy* 57: 526–539. doi:10.1016/j.landusepol.2016.06.013.

Schmidt, J. C. 2011. "What Is a Problem? on Problem-oriented Interdisciplinarity." *Poiesis & Praxis: International Journal of Ethics of Science and Technology Assessment* 7 (4): 249–274. doi:10.1007/s10202-011-0091-0.

Scholz, R. W., A. Spoerri, and D. J. Lang. 2008. "Problem Structuring for Transitions: The Case of Swiss Waste Management." *Futures* 41: 171–181. doi:10.1016/j.futures.2008.09.013.

Scholz, R. W., and G. Steiner. 2015. "Transdisciplinarity at the Crossroads." *Sustainability Science* 10 (4): 521–526. doi:10.1007/s11625-015-0338-0.

Steurer, R. 2010. "Sustainable Development as a Governance Reform Agenda: Principles and Challenges." In *Nachhaltigkeit regieren: Eine Bilanz zu Governance-Prinzipien und -Praktiken*, edited by R. Steurer and R. Trattnigg, 33–52. München: Oekom.

Stirling, A. 2008. "Opening Up' and 'closing Down': Power, Participation, and Pluralism in the Social Appraisal of Technology." *Science, Technology, & Human Values* 33 (2): 262–294. doi:10.1177/0162243907311265.

Strohschneider, P. 2014. "Zur Politik der Transformativen Wissenschaft." In *Die Verfassung des Politischen*, edited by A. Brodocz, D. Herrmann, R. Schmidt, D. Schulz, and J. S. Wessel, 175–192. Wiesbaden: Springer.

Ukowitz, M. 2014. "Auf dem Weg zu einer Theorie transdisziplinärer Forschung." *GAIA – Ecological Perspectives for Science and Society* 23 (1): 19–22. doi:10.14512/gaia.23.1.7.

Wagenaar, H. 2011. *Meaning in Action: Interpretation and Dialogue in Policy Analysis*. Armonk, N.Y: M.E. Sharpe.

Wittmayer, J. M., and N. Schäpke. 2014. "Action, Research and Participation: Roles of Researchers in Sustainability Transitions." *Sustainability Science* 9 (4): 483–496. doi:10.1007/s11625-014-0258-4.

Yanow, D. 2009. "Ways of Knowing: Passionate Humility and Reflective Practice in Research and Management." *The American Review of Public Administration* 39 (6): 579–601. doi:10.1177/0275074009340049.

Yanow, D., and P. Schwartz-Shea, eds. 2006. *Interpretation and Method: Empirical Research Methods and the Interpretive Turn*. Armonk, NY: M.E. Sharpe.

Zierhofer, W., and P. Burger. 2007. "Disentangling Transdisciplinarity: An Analysis of Knowledge Integration in Problem-Oriented Research." *Science Studies* 20 (1): 51–74.

Knowledge Decolonization *à la* Grounded Theory: Control Juggling in Research Situations

Maria De Eguia Huerta

ABSTRACT

Knowledge production is not free of political connotations. The researcher defines and moulds the research situation in which she will be gathering the data. Simultaneously, she will be also conditioned by the ways the situation is constructing her as a researcher. I shall elucidate some of aspects that influence how research situations are constructed based on the examples of my own empirical work. I will show some of the multiple negotiations contained in the process, influenced by the fact that I was a Spanish urban young woman doing research in a rural region of Bolivia for the German international cooperation for development (DED). Two sorts of control will be outlined and an argument for control juggling within a decolonial move of humanizing research will be sketched. I shall argue that Kathy Charmaz's constructivist Grounded Theory is flexible and systematic enough to make space for other ways of knowing, to grasp these subtle multifaceted processes of constructing a research situation while facilitating the necessary reflexivity for transformation purposes, contributing to a wider project of decolonizing knowledge production.

1. Introduction: Dynamics of Knowledge Production

There is an urgency of 'moving away from conventional Western modes of knowing in general in order to make room for other types of knowledge and experience' (Escobar 2012, 216). The current dynamics of knowledge production convert the South[1] into the excavation site for discovering and extracting data, to be then appropriated by the North for theorization and distribution purposes (Connell 2007; Tuhiwai Smith 2012). Global power relations are in this way further enforced through the axes of knowledge, research and imperialism by establishing the positional superiority of western knowledge in front of the Other (Tuhiwai Smith 2012).

For more than 500 years multiple forms of colonization by the North took place in the South, posing the base for domination through, amongst others, bodily and epistemic violence(s) justified through discourses on (superior) 'race' and 'culture' (Paredes Carvajal 2018). In doing so, 'through the "objectivity" of observation and science, European penetration into other lands [was] legitimised' (Loomba 2005, 57). Thereby, the epistemic hegemony as a form of domination shall be understood as one that is 'exerted not by force, nor even necessarily by active persuasion, but by a more subtle and inclusive power' (Ashcroft, Griffiths and Tiffin 1998, 116).

The methodological choices we make may contribute to challenging global power dynamics by critically questioning theoretical and practical 'assumptions of universality and attempting to re-centre attention on people, places, subjectivities and identities' (McEwan 2009, 103–104). A precondition for global social justice is global cognitive justice, which is based on the construction, acknowledgement and validation of knowledge and ways of knowing 'developed by social groups as part of their resistance against the systematic injustices and oppressions caused by capitalism, colonialism and patriarchy' (De Sousa Santos 2014, x). These include 'praxis', which is human being's 'reflection and action' upon the world 'to transform it', and which is necessary for the 'oppressed' to free themselves from the oppressor (Freire 2012, 40).

The present article approaches power relations within global dynamics of knowledge production while it explores the decolonizing potential of Charmaz's version of the Grounded Theory (GT) methodology arguing that it is a methodological approach that makes space for diverse ways of knowing. An autoethnographic component crosses the article. It is a methodological choice that relies on a longstanding feminist research tradition that puts situatedness (Haraway 1988) at the core and makes 'the personal' 'political' (Hall 2018) by unveiling power inequalities in relationships amongst human beings, while encouraging constant self-reflection (Ackerly 2008) as a keystone of the research process. In the present article a decolonial approach on global dynamics of knowledge production has been crossed with a feminist approach in order to methodologically explore both through the GT.[2]

First, I shall introduce the GT and possible contributions towards knowledge decolonisation. After that, specific aspects of what I will present as research situations will be sketched, based on practical examples of my empirical work, in the context of international cooperation for development in Germany and Bolivia.

The research this article refers to was about ways of conceiving human wellbeing in the light of the human capability approach and its possible connection to notions of gender equality. The human capability approach frames debates on local and universal ways of conceiving and measuring human wellbeing – e.g. in development programmes – (Sen 1993; Nussbaum 2000, 2011; Robeyns 2005). I explored the dynamics of gender equality promotion in the spaces of international cooperation for development against this background. A critical post-development stance brought me to pay more attention to the non-institutional discourses of those participating in this complex net of relations.

The empirical work phases of this research took place in Berlin and Bonn, Germany, and in diverse communities of La Chiquitanía and La Paz, Bolivia, between 2009 and 2011. I mostly interacted with 3 groups of people. First, the German development helpers – as their organisation called them – who were being trained in Germany by the DED (Deutscher Entwicklungsdienst),[3] and were about to leave for a few years to work in their assigned partner countries, in areas such as 'water resources management'.[4] Second, the Bolivian agricultural technicians[5] of a local organisation, which is typically designated as partner organisation, and which hosts development helpers that are supposed to assess them. Third, the agro-leaders[6] (as they called themselves), participants of a sustainable agricultural development project from the indigenous communities of the low-lands of Bolivia, who worked closely with the local technicians and the German development helpers. The discourses and experiences of these actors were mapped with a specific focus on gender politics from a postdevelopment[7] perspective (for more detail see De Eguia Huerta 2017).

Following some practical examples based on this research, references to epistemic violence and humanizing research will be made and two sorts of control will be sketched, while pleading for a juggling effort. A final word on research situations and further research questions will close this article.

2. Appreciating Charmaz's Grounded Theory: A Decolonizing Potential

Grounded Theory (GT) is a research methodology of an inductive nature: it is exploratory and aims at generating theory that arises from the data itself. It asks the researcher to constantly compare and follow some simple and systematic steps – such as gathering data, coding, writing memos and doing theoretical sampling – to construct from data an emerging theory. Its origins are to be found in the publication of *The Discovery of Grounded Theory* of Barney Glaser and Anselm Strauss in 1967. In the 1960s in the USA, quantitative methods predominated in universities and other institutions

(Charmaz 2006, 4). In this context the sociologists Barney Glaser and Anselm Strauss 'made a cutting-edge statement because it contested notions of methodological consensus and offered systematic strategies for qualitative research practice' (Charmaz 2006, 5).

At its beginning it was the 'notion of generating new theory from the data, as opposed to testing existing theory' (Birks and Mills 2015, 2) that made GT popular and gain many interested scholars who further developed it in different directions. Some of the most well-known second-generation grounded theorists are Juliet Corbin, Kathy Charmaz and Adele E. Clarke and Janice Morse (Birks and Mills 2015, 2; Mey and Mruck 2011). After GT as a 'research style' (ibid., 16) was born diverse propositions were developed so that we should be talking about 'GT Methodologies in plural' (ibid., 12). Additionally, to the GT tradition that has evolved from Glaser and the one from Strauss (and Corbin), Charmaz developed GT further and adapted it becoming 'the most important representative for a constructivist GT methodology approach' (ibid., 18).

Charmaz's GT's proposal, to which this article refers to, calls for staying creative, open and flexible throughout the whole process of research. Charmaz follows Glaser's principle that 'all is data' (Charmaz 2006, 16) while for her, 'data is constructed through our observations and interactions, as well as diverse sort of materials that we may collect' (Charmaz 2006, 3). In terms of knowledge production, Charmaz strongly distances herself from the 'classic' GT of Glaser and Strauss. For unlike the founders, she argues that the researcher is part of the world she approaches to be studied and to get data from, since 'we *construct* our grounded theories through our past and present involvements and interactions with people, perspectives, and research practices' (Charmaz 2006, 10). We must distinguish the philosophical methodological pillars of the different GT versions from the methodological practices that may derive from it. For this often depends on how the researcher applies them, e.g. Charmaz's GT version is not automatically participatory, different researchers may make an unalike use enhancing or reducing levels of participation.[8]

Epistemology as the study of knowledge is closely related to the undertaking of decolonising projects (Prada Alcoreza 2013). Epistemological pluralism refers to a real co-existence of diverse ways of knowing, which are equal in value, which exist in constant horizontal conversation (Prada Alcoreza 2013), are complementary (Vilsmaier, Brandner, and Engbers 2017) and give 'common people [...] sufficient control over the generation of new knowledge [...] encourage[ing] the remaking of knowledge and science for the benefit of the masses victimized by power' (Fals-Borda 1991, 147). This contrasts the hierarchy that impregnates certain positions of 'multiculturality', in which the 'modern culture' makes 'democratic concessions to other native cultures' (Prada Alcoreza 2013, 14).

Intellectual work takes place in multiple ways. However, knowledge produced in the South is often questioned and marginalised by the dominant North American and European metropoles (Connell 2007). Knowledge and forms of knowing of non-Western origin are often perceived as folkloric, following the imaginary that the North possesses scientific knowledge and the South possesses rich culture, being both constructed as static, opposed and incompatible (Mohanty 1984; Escobar 2010, 2012; Tuhiwai Smith 2012).

Coloniality differs from colonialism inasmuch as 'coloniality survives colonialism' through 'long-standing patterns of power' which are 'alive in books, in the criteria for academic performance, in cultural patterns, in common sense, in the self-image of peoples' (Maldonad-Torres 2007, 243). GT precisely turns to these everyday experiences of human beings, in which we find components of macrostructural global injustices, and supports the researcher in her task of unveiling and decoding them. GT permits an open research design and execution process that pushes the researcher to 'uncover contradictions, deconstruct dominant discourses and thereby contribute to the decolonization of knowledge' (Schultz 2014, 92).

A 'decolonial attitude' means that one has the 'responsibility and the willingness' to consider 'the points of view of those whose very existence is questioned and produced as insignificant' (Maldonad-Torres 2007, 262). GT is a suitable methodological choice for those willing to include these voices. For 'you let the obdurate empirical world speak to you, you listen, take notes, write memos to yourself, form writing groups. No hierarchy, the social theorists are not privileged. In the world of GT anybody can be a theorist' (Denzin 2007, 454). GT's core of witnessing the process of emergence of theory does correspond with a decolonializing 'attitude' that 'requires detachment and wonder' (Maldonad-Torres 2007, 262).

As Julieta Paredes Carvajal (2013) puts it: '[. . .] There is no such thing that one produces ideas from her bellybutton, the bellybutton connects us always with the other, who is the mother and from this initial relation, we are always constructing and creating in dialog and within a community'. Research relationships are based on co-creation. Thereby, these relationships are not always of collaborative-horizontal nature. Power (re)distribution issues taint these interactions, making a decolonial move urgently needed.

Charmaz's GT proposal can foster epistemological pluralism by staying radically open when standing in front of data, and broadening our filter when deciding what is data. GT allows us to encounter ideas or perspectives, amongst others, that otherwise we would have never been able to see or hear. An example from my research is the notion of a 'vida tranquila' (De Eguia Huerta 2018), which literally would be translated as a 'quiet' or 'calm' life. However, it was a much richer notion challenging ideas of development in a developmentalist context that turned out to be part of the skeleton of the categories that arose.

Under certain conditions, GT allows the researcher to position herself in a place from which she may be able to perceive these other ways of knowing. This is a decisive step towards transforming collaborative research relationships and research more towards a human intellectual activity nurtured by reflexivity. GT asks the researcher through the writing of memos, for (self-)reflexivity but also transparency, both, key aspects to exercise self-control and deconstruct the mythos of an impartial researcher contributing to a specific legitimated knowledge, which becomes indisputable.

Charmaz's (2006) GT challenges the researcher constantly confronting her with her own prejudices and assumptions, taking for granted that these will be influencing the analytical direction of the study. During my empirical work (De Eguia Huerta 2017), as with the whole research process, I followed the principles of Charmaz's GT. Taking my own interventions as part of data, constantly analysing my position[9] and the effects this could have on the research, made it possible for me to sketch some of the ways which my very presence was affecting the research situations and the whole research process.

GT is primarily suitable for qualitative research, which may be a smoother gate to enter the decolonizing path. Mruck and Mey (2007) stated that 'in times of social change and globalisation, the limitations of quantitative methods are evident since, by definition, they are primarily useful with regard to theories or hypotheses derived from existing and established theories' (p.515). Within (pre-)established theories colonial power can easily keep on circulating. In this sense, using Charmaz's GT, as any other methodology, is not only a methodological choice, but a political one, too. For 'knowing' and 'inquiry' are 'always moral and political' (Denzin 2007, 458 and 466).

GT fosters and demands high levels of reflexivity during the complete process of researching. If we attempt to demystify those privilege-fostering ideas about knowledge production, we must not stop at making visible unjust power relations at a macro level. For knowledge production is just one more human activity, and as each other activity, it functions based on interrelations at a micro level. The idea of research situations that are constructed based on specific negotiations, which I shall develop, does situate itself there.

Next to many feminist authors who have intensively worked on positionality (Harding 1991; Carotenuto, Jambresic Kirin, and Prlenda 2014 or Haraway 1988, amongst others), one of Charmaz's main contributions has been to make clear the place of the researcher: 'My emphasis on constructivism loosens GT from its objectivist foundations and brings the grounded theorist into the research situation and process of inquiry. We stand *within* our research process rather than above, before, or outside it' (Charmaz 2014, 321). Furthermore, Charmaz has pleaded for an engaged attitude also described by Vilsmaier, Brandner and Engbers as an 'interest, openness, empathy and willingness to learn' (2017, 175). While this has been explained in theory, there is little concrete explanation of how these dynamics look in the very concrete situations in which researchers practise the 'artisan craft of research with the other' (Espitia Vásquez 2008, 109). The following section contributes to this gap.

3. Dynamic Aspects of Research Situations

According to Charmaz (2006), 'an interview is contextual and negotiated' (27). Each setting of a conversation of any sort – including formal interviews – within the frame of research practices is a constructed one, in which 'in addition to the dynamics of power and professional status, gender, race and age may affect the direction and content of the interviews' (Charmaz 2006, 27) and work as axis of the negotiations that take place during the process of moulding a research situation.

A research situation, such as performing participant observation or an interview, does not begin with the first question, but with arriving to the place. The situation of researching *with* the other may have diverse intensity levels in terms of openness and active collaboration. In what follows, I shall refer to 'conversations', including diverse sorts of encounters and talks with one or more people, which varied from semi-structured to open, and lasting from 30 minutes to a few hours. The strong diversity of research situations I will refer to also comprises diverse roles in terms of collaborative research. In some of those situations people involved in the situation would adopt an 'informant' role, while in others they were very active co-researchers. For practical reasons, I will refer to them as 'research participants' but I consider them to be co-definers of the research situation.

There is no sensorial or rational space for separating context from specific research practices, for the context is always part of the dynamic research situation. The context pushes research participants and the researcher to deal with diverse issues within a research situation. And these negotiations, in turn, contour the setting that is prompting the research situation. These conditions change during the researching practice and may be dealt with anew, for instance, if a very sensitive issue arises in the conversation.

The careful moves towards spoken or unspoken agreements I experienced with the research participants depended on diverse aspects, some of which I shall explain in what follows. They moulded specific research situations, which existence and importance I stress as part of an analytical decolonizing movement. These are some of the results of a research endeavour that took several years (De Eguia Huerta 2017), which here are only briefly mentioned.

1) The stage of the research and the (anti-)positivist battlefield. Some of the people I interacted with during my empirical work (development helpers or top managers of the organisation, amongst others) questioned my research steps (such as going back to the field while coding was not completed) and me being a 'good scientist' more at the beginning of my research than in the advanced stages of it – where I was, for instance, searching for specific parameters within discourse structures to saturate a category. GT's very open approach to the issues that I seemed to be 'waiting for to emerge' was confusing for some of the research participants, who were expecting individual structured and close interviews. Ideas of positivist science based on (pre-designed) hypothesis to be proved, intertwined with other issues of privilege interacted and resulted into palpable differences in the way people interacted with me and my research project. The further from the academic world people were situated, the less confrontational they were in questioning me and my approach. Thus German (institutionally) educated development helpers tended to react more disturbed in front of collaborative open-ended procedures than Bolivian agro-leaders from the indigenous communities.

2) Being a mother and being married. The agro-leaders in Bolivia did directly ask me if I was married and if I had children quite at the beginning of our conversations. Once this was clear they moved on to other issues, except for those women who expressed to be worried about why at that time I 'only' had two children and insisted I should put more effort into my family plans while they argued about the advantages of having more children. With the local technicians (men) this happened to a lesser extent. Some of the development helpers from Germany changed the way they approached me as a ('very' or 'too') young woman until they found out I had children. They then offered me more horizontality and complicity (e.g. an invitation to a specific social event would be then available for me). I learnt with surprise that in Germany, in terms of privileges, being a mother, generally, compensated for my being considered too young, and that power relations circulated once more through the intimate veins of gender normativity, and here more specific, through imaginaries about motherhood. In Bolivia, I was not seen as too young, but as having too few children, especially among the agro-leaders. Thus having 'only' two children put me in a situation of being openly questioned by the people I met.

3) The ruling language. Even though I understand and speak German it is not my mother tongue – Spanish and Catalan are. The conversations in German needed some language negotiations, at times mentally tiring but often very enriching, too: often the research participants took for granted I did not understand certain words or expressions and clarified them for me, which gave me extra amounts of information.

This was different in Bolivia. For the technicians and the agro-leaders I spent time with, defining themselves as indigenous, mostly used Spanish, the language of the colonizer, as the main communication language at home and outside their homes. Still, two versions of the Spanish language came together: the one I brought with me and the one people in *La Chiquitanía* spoke. The second included many words that people had been adopting and including from the ancestral language *Besiro* and others that had been transformed from Spanish over the years. Thus, despite the research participants and me sharing the language, expressions, vocabulary and accent reminded me that these were two versions of 'Spanish', that represented two sides of a dramatic and violent history of colonization. This made me feel very uncomfortable at the beginning. People tried to speak 'more elegantly' – as one woman said – in front of me. This changed after a few hours, for it was too exhausting. Later, some people even laughed at the way I spoke Spanish.

The language was also constantly an issue to deal with between technicians and agro-leaders, for other issues, such as educational level, played an important role.

4) The North and the South. During my conversations with the agro-leaders and the technicians in Bolivia I was perceived as belonging to the affluent geopolitical North, for I was not only put in relation to Spain, through the way I look and the way I talked, but I was also put in relation to Germany, because of the institutional arrangements. This forced me to negotiate many issues, such as being used as 'a megaphone', as Mruck and Mey (2007) have named the situation when people involved in an interview try to communicate their needs or perspectives to others through the interviewee (522). For the agro-leaders often had the hope that I would report about their wishes to continue the project to the DED or that I even had some sort of influence, which I did not have[10].

On the contrary, during my empirical work stays in Germany, I was approached as the 'southern girl'. In some cases, more than in others, I had to reposition myself, for 'the Mediterranean' or 'the South' – even if it is 'only' Southern Europe – does imply some meanings that put me in a position in which I did not want to be (e.g. I actively refused being someone's prolongation of a nice holiday, as some development helpers were rather interested in talking to me about their last vacation in my homeland or about how great in the place where I come from life must be, because the sun shines and people are that joyful). Binary, stereotyped North-South imaginaries took the room and I had to work hard to position myself and my research and re-conduct the conversation. I used diverse strategies, such as improvising a humoristic comment that questioned those assumptions based on stereotypes or posing a specific question to introduce an issue that made the conversation content wise dense and pushing us both to dialogue based on more technical or theoretical issues.

5) Too much and too little of gender. I encountered people while doing empirical research who referred to me as the 'gender-woman'. To take the label of 'gender' away from me demanded some effort. On the contrary, some of the 'gender experts' I interviewed were confused about me talking in 'human well-being' terms and even tried to convince me about the importance of approaching gender issues staying 'loyal' to gender. They found my approach weakened a 'truly' gender agenda because of the intertwining with issues of human well-being.

6) Being familiar and being a stranger. Mruck and Mey have stated, that 'both being familiar and being a stranger influences data collection (and data analysis)' (2007, 523), something which functions depending on diverse axes of variations, some of which I have just listed. On one side, 'the more researchers and participants belong to similar cultures, the more interviewers may pre-suppose concepts and values as shared, and therefore fail to attend to elaboration during the interview, so that even central topics will be obscured or lost' (Mruck and Mey 2007, 523), and the researcher loses the opportunity of accessing data regarding meanings, as I have previously mentioned when referring to language. 'Conversely', so Mruck and Mey (2007), 'intimate knowledge will hardly be given to strangers, so familiarity will provide other insights than in the case of a researcher unfamiliar with a person or an issue' (523).

The being familiar and being a stranger is a fluid status in constant negotiation, and at times, as I experienced it, in a matter of a few minutes, depending on the issue or on the presence of someone else, I stopped being a stranger and became familiar or vice versa. For instance, not being German was a source of familiarity for the Bolivian technicians when the talk was about the German cooperation for development as an institution. However, on other occasions, the fact that I presented myself as a woman, made some of the technicians treat me less familiarly when female agro-leaders were around, as if our sex had put us in two different sacks. And even this changed at times, for the fact that I had a university degree like most of the technicians, on certain occasions, seemed to play a stronger role than our sex.

7) (In)formal situations. A further aspect of research situations is their degree of formality. The research situations I was involved in had a certain degree of formality or informality – as part of a continuum – defined, amongst others, by the place in which I interacted with people, the way I presented myself, having the recorder on or not, the presence of other (un)expected people, and the point in time – the first conversation was always rather formal, and the last ones were mostly informal. Coding and constant comparison, amongst others, allowed me to become aware of these patterns and include them in my analytical work, as I shall show in a few lines.

So far, I have mentioned some dynamic and interconnected aspects of research situations that I have explored based on my research. These aspects mostly belong to the process of building relationships with the research participants. They were at the same time moulding specific research steps, for the way participants get involved while data gathering, for instance, is already data. In what follows, I shall contribute with some thoughts on decolonizing research by humanising it before arguing for control juggling when constructing research situations.

4. Epistemic Violence and Decolonizing by Humanizing

Analysing the (in)formality degree of a research situation, within a decolonial move, only gets transformative if we reflect about the specificities of the epistemic violence involved in institutional-formal settings.

While involved in spiral-structured procedures of going back to the field after analysing data I stumbled upon a code that started to gain strength, named 'two-chair and one recorder situation'. I could analyse some aspects of institutionally formal planned research situations tainted with epistemic violence, in which people were forced by my invitation to communicate in a particular place, time of the day, sort of language and even through a specific mechanism of asking-answering. Becoming aware of it through the GT's inductive coding, constant comparing and memos writing, made it possible for me to search for an alternative and actively alter the process. For instance, I put effort into travelling alone to the agro-leaders instead of being always accompanied by the technicians or began to propose walking interviews instead of static ones, hoping for a communicational momentum to come to life. After that, I had conversations while travelling in a truck, working in the fields or having a walk (Kühl 2016). A decolonial move within situations of collaborative research is about which knowledges we praise and which we disdain, but also about the how, about the ways we construct research situations.

Non-formal conversations had a much stronger 'now-it-is-my-turn effect' than the formal ones. This was the way I coded it when research participants posed some questions for me, too. Proactively constructing research situations of horizontal nature is not done in one move. Multiple small decisions will be taken by the researcher during the interactions within the situation that determine to which extent control will or will not be shared.

I wrote many memos about the concrete meaning of staying at eye-level – as much as possible despite strong issues of power due to age, ethnicity, language, etc. Feeling unsure about constructing research situations that enabled just human relationships started to paralyze me. At one point, I took the decision of translating that intention into just staying human. Hence to think, feel and act as Maria and not as a researcher. In one specific issue this became very liberating.

Two weeks before my departure to Bolivia my beloved grandmother passed away. Coming from a family in which mourning is very much a collective process, and not being able to participate in this for being away, there were some more emotional moments, which I did share with the people I spent time with. The death of my grandmother was, obviously, out of my control. Still, it did construct the specific research situations in which I was involved, for sharing my feelings also shaped the relationship with some participants. I took the decision to consciously experience research as an activity which is intertwined with other aspects of (my) life. I find it worth to further think of humanizing research as one of the moves to decolonize knowledge production.

In the sense of Freire's work on 'humanizing education' (2012) research practises need to be humanized, too. And in this sense, GT accompanied me through the process of finding a methodological, ethical and emotional comfortable reflexive place to stay from which to try to do so. The initial interaction with other human beings is determining for building just human relationships. Thus we can at this point say, that a methodological framework that strengthens us in reflexive practices of constructing research situations from the very beginning enwombs a potential for a decolonial move – while we broaden our understanding about when research starts and disrupt a linear conception of doing research and then reflect about it.

It is my aim to contribute to deconstruct the figure of the researcher and promote a more human understanding of her. Thus, seeing her as a human being with personal experiences that will shape the research situations, too. For personal life does not stop while one researches. Therefore, it is senseless and fallacious to detach research from personal life, one human being from others related to her, one set of data from the diverse dynamic aspects of a research situation that have conditioned the one who gathered them. Decolonizing research situations by humanizing involves an embrace of informal situations, which despite containing power relations, too, offer broader space for epistemic pluralism and which in turn, escape control of many sorts.

5. Control Juggling

I did have this one picture of my children in my wallet. During a conversation, I decided to show it to one Bolivian technician, after he had posed many questions to me. This one simple picture played a role in terms of constructing situations. It kindly opened the door for me to many people in a more human way. It was one of the sweetest and softest entrances into research situations, which was at the same time humanizing and carrying consequences in terms of control. Often, one of the technicians that would travel with me to a community would tell the agro-leaders that I had two children and a picture of them with me – thus other labels, such as the institutional one, were weakened by his way of introducing me. Inevitably, I was asked for the picture.

The picture became a power and control-sharing utensil. For the age of my two children were easy to guess in that picture. This raised in every community similar questions or even direct comments that disapproved me leaving the 'too young' children with their dad. This photo of my personal life constructed the research situation positioning me in a very different place than without it. For even though the share of personal information through the picture was a voluntary and conscious decision, the use that the people I interacted with would make of it was not under my control.

Control share implies that one party renounces to have all privileges for the sake of sharing them. The endeavour of decolonizing collaborative knowledge production involves to justly deal with research situations in its construction process. It is a micro level of macro structures that inject power issues into very subtle specificities of a situation. GT enables the researcher to perceive, consider, analyse and reflect about her role in those situations that she constructs and in which she is constructed as a researcher. Through reflexivity the researcher can cultivate sensitivity about her privileges and decide in which ways will she share the control of this research situation.

Certain circumstances or events, predictably and unpredictably, may put the researcher in a position in which (re)negotiation is needed, and in any case denying her the absolute possibility to have an overall control. Having the possibility to decide to which extent one shares control of the research situation requires a starting point of having that control. Thus, here one could ask to which extent other methodological traditions, such as the Latin-American Participatory Action Research (e.g. Fals-Borda 1991) would be more suitable to look at in order to easier overcome or avoid control distribution issues. Going into detail on this conversation between PAR and GT could be developed in a future work. For it is beyond the scope of this article and deviates from the original intention of exploring a decolonizing potential of Charmaz's GT proposal.

On the one side, a decolonizing move within knowledge production requires the researcher to abandon the type of control that is based on abusing from power. On the other side, aspirations of collaboratively elaborated knowledge go in hand with the need to systematize and make reflection possible and translatable to contribute to the understanding and tackling of the most pressing global issues affecting the Earth and its inhabitants. For this to be technically possible some sort of control needs to be in hand of the (co)researcher(s) and the research project. These are two sorts of control.

GT is a good companion for those seeking to be 'an insider who wishes to deconstruct from within the Western academy and its positivist epistemologies' (Denzin 2007, 457). As Charmaz (2014) has argued, for research to have a socially just outcome, knowledge must be co-created with research participants, while those very relationships must be examined, too. Collaborative research situations may hide power relations and this first sort of negative control serves to the prolongation of privileges around axes of sex, gender, religion, race, age or educational background, amongst others.

The positive side of control is about perceiving and including as much as possible from the research situation we are in, embracing data in a holistic way. By broadening the construction of research situations and not delimiting which kind of information, experiences, voices we should include in our research, we confront a huge amount and richness of data that must not be lost. GT allows us to receive, hold and absorb that valuable data. Not only novel results come out of this process. The positive control is the one permitting us to notice and reflect upon the negative sort of control, the one that we will need to handle about by doing one or several decolonial moves within the process of constructing research situations.

Following the GT does not guarantee that the researcher will be able to give up the privileges and power-related control, while creatively keeping the systematic and all-capturing control. For this to be so, there is a chain of big and little decisions that she will be taking throughout the whole process. It is a constant juggling of control, that the researcher will have to handle standing on the grounds of her humanity. This is the political agenda of humanizing research within the methodological, and this needs to be further researched and put at the centre of the debate. The GT offers an enquiry style that makes space for a decolonising and humanising turn of research practices.

6. Conclusions and Further Research on the Politics of Methodology

I follow Mignolo in his proposal of taking 'decoloniality as a disrupter in the academy' (Mignolo and Walsh 2018, 106) and do agree with him when he argues that 'there is not [...] master plan for decoloniality' and that the question about the meaning of decolonizing is not abstractly answerable but 'has to be answered by looking at other W questions: Who is doing it, where, why and how?'

(Mignolo and Walsh 2018, 108). Inspired by Mignolo's work on modernity, coloniality and decoloniality as well as other works on decolonial feminism less known in the academic circles, such as Paredes work (2013, 2018), in this article I have extended an invitation to take a methodological journey inspired in Charmaz's GT, that poses deep methodological reflections which we urgently need to further discuss and broaden.

The promises that Eurocentric knowledge production has made under the name of science (Connell 2007; Escobar 2012) must be questioned to create space for other ways of knowing. We must embrace the 'principle of incompleteness of all knowledges' as 'the precondition for epistemological dialogues and debates among different knowledges' (De Sousa Santos 2014, 189) to overcome fallacious knowledges (hegemonic) dichotomies of North-South or 'academic and popular' (Fals-Borda and Mora-Osejo 2003, 36), amongst others. A methodological pathway is needed to create space for 'dissident voices' (Hooks 2003, 187) that challenge domination structures of 'imperialist white-supremacist capitalist patriarchie[s]' (ibid., 10). A postcolonial move needs 'resistant methods' (Castro Varela and Dhawan 2005, 138) that 'confront the logic of the monoculture of scientific knowledge' (De Sousa Santos 2014, 188) in the arena of methodological debates.

Research situations are constructed in an unceasing, hybrid and dynamic game of intertwined identities, experiences, labels, perceived commonalities and differences. Dynamic aspects of the construction of a research situation push the researcher to constant negotiations. In each of those negotiations a potential methodological and ethico-political choice will be facing the researcher. This will be a decolonizing momentum. Ideally, skilfully juggling of control sorts by the researcher will make space for a situation of collaborative research, making a step towards power (re)distribution, while a holistic capturing sort of control will be preserved.

Charmaz's GT proposition makes space for complexity and reflexivity, it is appropriate to combine different ways of knowing and to acknowledge other ways of accessing, understanding and producing knowledge. It enables researchers to move differently in the parameters of socio-political research while it redefines what knowledge (production) may be.

Despite these contributions, in a world in which knowledge production and knowledge possession regulates politics and economics in such a strong way, to openly debate about these taken-for-granted assumptions when we refer to knowledge production remains an unfinished task. This is not a matter to be solved with one fast move, nor does it mean we may better renounce to research to avoid its problematic consequences. On the contrary, brave people are needed in the search for collective created knowledge.

Emotional and intellectual energy should be put into the further methodological evolvement of the GT and any other methodological impulse that enables us to adroitly juggle control of all sorts, and at all stages of our research endeavour. A dialogue between GT and other transformative methodological proposals from around the globe should continue or even get started to answer those urgent questions about the scope and dimensions of the politics of methodology when it comes to global social justice endeavours of research practices.

Notes

1. I use 'North' and 'South' in a metaphorical sense using the political value of it purposed by Mohanty (2003).
2. I thank an anonymous reviewer for helping me clarify this linkage.
3. Later on the GIZ, Deutsche Gesellschaft für Internationale Zusammenarbeit.
4. The 23 individual and group interviews were from very open to semi-structured and took place with men and women from 20 to 50 years old, often from a middle-class background, mostly from Germany. The reasons for being involved in this institution were varied and their problematic position – in which they were partly structurally 'trapped' – has been analysed in depth elsewhere (De Eguia Huerta 2017). Broad data was gathered since I lived in the facilities while they were trained with them and beyond the interviews, many varied sorts of interactions took place.

5. They were 1 woman doing coordinating tasks in the city and 5 men doing the field work in the indigenous communities, to which they partly belonged. Very broad data was gathered through very open interviews and many interactions during the weeks I shared with them.
6. 14 agro-leaders from 4 communities were interviewed and visited in their plot and/or house. I often spent some days with them and accompanied them to the field. Big amounts of data were gathered during these interactions.
7. Postdevelopment is a 'school of thought in development theory which is fundamentally critical of the very idea of "development" and promotes alternative ways of thinking and acting beyond this idea' (Ziai 2012).
8. I thank Bob Dick for his very illuminating comments helping to enrich this article.
9. I also followed the same procedure for the people I interacted with, reflecting about their background, possible interests or perceptions they could have, amongst others, that could play a role in the way the constructed the research situations in which we were involved. This is however not the point I am trying to make in this article: Charmaz's Gt has the potential to accompany the researcher in a decolonial move and make space for diverse ways of knowing, amongst others, because it pays attention to positionality and power dynamics at the micro level of knowledge production.
10. I have elsewhere explained my role as an external researcher during my field work phases with the DED, my institutional disconnection, and the challenge of mostly being only tolerated, and in any case maintained far away from any power over decisions (De Eguia Huerta 2017).

Disclosure statement

No potential conflict of interest was reported by the author(s).

References

Ackerly, B. 2008. "Feminist Methodological Reflection." In *Qualitative Methods in International Relations. Research Methods Series*, edited by A. Klotz and D. Prakash, 28–42. London: Palgrave Macmillan.

Ashcroft, B., G. Griffiths, and H. Tiffin. 1998. *Key Concepts in Post-colonial Studies*. London: Routledge.

Birks, M., and J. Mills. 2015. *Grounded Theory – a Practical Guide. 2nd Ed.* London: SAGE Publications Ltd.

Carotenuto, S., R. Jambresic Kirin, and S. Prlenda. 2014. "Editorial Introduction." In *Feminist Critique of Knowledge Production*, edited by S. Carotenuto, R. Jambresic Kirin, and S. A. Prlenda, 5–20. Napoli: University Press.

Castro Varela, M., and N. Dhawan. 2005. *Postkoloniale Theorie: Eine kritische Einführung*. Bielefeld: transcript Verlag.

Charmaz, K. 2006. *Constructing Grounded Theory: A Practical Guide through Qualitative Analysis*. London: Sage Publications.

Charmaz, K. 2014. *Constructing Grounded Theory*. 2nd ed. London: Sage Publications.

Connell, R. 2007. *Southern Theory: The Global Dynamics of Knowledge in Social Science*. Cambridge: Polity Press.

De Eguia Huerta, M. 2017. "How Can the Human Capability Approach Contribute to Gender Mainstreaming? A Post Development Perspective on the Dynamics of Gender Equality Promotion in the International Cooperation for Development." In *Perspectives on Development*, edited by P. Dannecker and U. Schultz. Münster: LIT Verlag.

De Eguia Huerta, M. 2018. *Vida Tranquila. Alternativen zur Entwicklung am Beispiel des Gender-Mainstreaming in Bolivien*. PERIPHERIE, 2-2018,290–302. doi:10.3224/peripherie.v38i2.09.

De Sousa Santos, B. 2014. *Epistemologies of the South: Justice against Epistemicide*. USA: Paradigm Publishers.

Denzin, N. 2007. "Grounded Theory and the Politics of Interpretation." In *The SAGE Handbook of Grounded Theory*, edited by A. Bryant and K. Charmaz, 454–471. London: SAGE Publications.

Escobar, A. 2010. "Una Minga para el postdesarrollo: lugar, medio ambiente y movimientos sociales en las transformaciones globales." *Programa Democracia y Transformación Global Universidad Nacional Mayor de San Marcos. Facultad de Ciencias Sociales Lima: Universidad Nacional Mayor de San Marcos*. Accessed 22 September 2018. http://www.unc.edu/~aescobar/text/esp/escobar.2010.UnaMinga.pdf

Escobar, A. 2012. "Encountering Development: The Making and Unmaking of the Third World." *E-reader Version (1995 Version with a Preface from 2012)*. Princeton and Oxford: Princeton University Press.

Espitia Vásquez, U. 2008. "Producción de conocimiento, prácticas intelectuales y reflexividad." *Nómadas. No.29*, 96–111. Bogotá, Colombia: Universidad Central. Accessed 11 December 2018. http://www.redalyc.org/pdf/1051/105112131008.pdf

Fals-Borda, O. 1991. "Remaking Knowledge." In *Action and Knowledge: Breaking the Monopoly with Participatory Action-Research*, edited by O. Fals-Borda and M. Anisur Rahman, 146–16. New York: Apex Press.

Fals-Borda, O., and L. E. Mora-Osejo. 2003. "Context and Diffusion of Knowledge. A Critique of Eurocentrism." *Action Research* 1 (1): 29–37. SAGE Publications, London, Thousand Oaks CA, New Delhi. doi:10.1177/14767503030011003.

Freire, P. 2012. *Pedagogía del Oprimido*. 2nd ed. Madrid: Siglo XXI de España Editores, S.A.

Hall, M. S. 2018. "The Personal Is Political: *Feminist Geographies Of/in Austerity*." *Geoforum*. doi:10.1016/j.geoforum.2018.04.010.

Haraway, D. 1988. "Situated Knowledges: The Science Question in Feminism and the Privilege of Partial Perspective." *Feminist Studies* 14 (3): 575–599. Published by: Feminist Studies, Inc. doi:10.2307/3178066.

Harding, S. 1991. *Whose Science? Whose Knowledge? Thinking from Women's Lives*. Ithaca: Cornell University Press.

Hooks, B. 2003. *Teaching Community: A Pedagogy of Hope*. New York: Routledge.

Kühl, J. 2016. "Walking Interviews als Methode zur Erhebung alltäglicher Raumproduktionen." *Europa Regional* 23.2015 (2): 35–48. Accessed 15 September 2018. http://nbn-resolving.de/urn:nbn:de:0168-ssoar-51685-8

Loomba, A. 2005. *Colonialism/Postcolonialism*. 2nd ed. Oxon: Routledge.

Maldonad-Torres, N. 2007. "On the Coloniality of Being." *Cultural Studies* 21 (2–3) March/ May: 240–270. Accessed 11 December 2018. http://www.decolonialtranslation.com/english/maldonado-on-the-coloniality-of-being.pdf

McEwan, C. 2009. *Postcolonialism and Development*. Abingdon: Routledge.

Mey, G., and K. Mruck. 2011. "Grounded-Theory-Methodologie." In *Grounded Theory Reader*, edited by G. Mey and K. Mruck. 2nd ed., 11–48. Wiesbaden: VS Verlag für Sozialwissenschaften.

Mignolo, W., and C. Walsh. 2018. *On Decoloniality: Concepts, Analytics, Praxis*. Durham and London: Duke University Press.

Mohanty, C. T. 1984. "Under Western Eyes. Feminist Scholarship and Colonial Discourses." *Boundary* Bd. 12, Nr. 3 bis Bd. 13, Nr. 1., S. 333–358. doi:10.2307/302821.

Mohanty, C. T. 2003 Winter. "'Under Western Eyes' Revisited: Feminist Solidarity through Anticapitalist Struggles." *Signs* 28 (2): 499–535. The University of Chigaco Press. Accessed 20 September 2018. http://www2.warwick.ac.uk/fac/arts/english/currentstudents/postgraduate/masters/modules/postcol_theory/mohanty_under_western_eyes_revisted.pdf

Mruck, K., and G. Mey. 2007. "Grounded Theory and Reflexivity." In *The SAGE Handbook of Grounded Theory*, edited by A. Bryant and K. Charmaz, 515–538. London: SAGE Publications.

Nussbaum, M. C. 2000. *Women and Human Development: The Capabilities Approach*. Cambridge: Cambridge University Press.

Nussbaum, M. C. 2011. *Creating Capabilities: The Human Development Approach*. Cambridge, Massachusetts: Belknap Press of Harvard University Press.

Paredes Carvajal, J. 2013. Accessed 12 December 2018. http://seminariodefeminismonuestroamericano.blogspot.com.es/2013/04/hilando-fino-desde-el-feminismo.html

Paredes Carvajal, J. 2018. *Descolonización del pensamiento occidental*. Instituto de Estudios Latinoamericanos,University of Stockholm. Accessed 12 December 2018. https://www.youtube.com/watch?v=SnJsgVWa8Lk

Prada Alcoreza, R. 2013. "Epistemología, pluralismo y descolonización." Accessed 20 April 2016. http://www.rebelion.org/docs/167277.pdf

Robeyns, I. 2005. "Sen's Capability Approach and Gender Inequality: Selecting Relevant Capabilities." In *Amartya Sen's Work and Ideas: A Gender Perspective*, edited by B. Agarwal, J. Humphries, and I. Robeyns, 63–94. Abingdon, Oxon: Routledge.

Schultz, U. 2014. "'Über Daten nachdenken': GT Studien in der entwicklungsbezogenen Forschung." In *Qualitative Methoden in der Entwicklungsforschung*, edited by P. Petra Dannecker and B. Englert, 75–93. Wien: Mandelbaum.

Sen, A. 1993. "Capability and Well-Being." In *The Quality of Life*, edited by M. C. Nussbaum and A. Sen, 30–53. Oxford: Clarendon Press.

Tuhiwai Smith, L. 2012. *Decolonizing Methodologies: Research and Indigenous Peoples*. 2nd ed. London: Zed Books.

Vilsmaier, U., V. Brandner, and M. Engbers. 2017. "Research In-between: The Constitutive Role of Cultural Differences in Transdisciplinarity." *Transdisciplinary Journal of Engineering & Science* 8: 169–179. doi:10.22545/2017/00093.

Ziai, A. 2012. "Post-Development." *InterAmerican Wiki: Terms - Concepts - Critical Perspectives*. Accessed 12 December 2018. www.uni-bielefeld.de/cias/wiki/p_Post_Development.html

Identity Politics: Participatory Research and Its Challenges Related to Social and Epistemic Control

Stefan Böschen, Martine Legris, Simon Pfersdorf and Bernd Carsten Stahl

ABSTRACT

Over the past 20 years, the participation of laypersons or representatives of civil society has become a guiding principle in processes of research and innovation. There is now a significant literature discussing collaboration between civil society organisations (CSOs) and researchers, with two interesting gaps. Firstly, the fact that research is mainly conducted within projects is often underestimated, although the format significantly frames knowledge production. Secondly, researchers and civil society organisations are closely related to their respective communities. We argue that this constellation – of project-related format, in combination with a strong relationship to communities – results in conflicts that express and lead to identity politics. The analysis is based on conceptual considerations as well as empirical findings, which were developed within the EC-funded CONSIDER project (2012–2015). It can be shown that identity politics is performed by socio-epistemic tactics, which are used to order the socially as well as epistemically hybrid space within projects. To explain differences in conflict intensity, we suggest the distinction between weakly tied and strongly tied identity politics. In sum, identity politics can be seen as one key element for social as well as epistemic control in transdisciplinary research projects.

1. Introduction

The participation of laypersons or representatives of civil society in processes of research and innovation has acquired the character of a guiding principle. A milestone within this debate was an early study by Epstein (1996), pointing at the blurring of boundaries between science and society. Thereby, quite different 'hybrid forums' of research (Callon, Lascoumes, and Barthe 2009) were evolving, indicating and provoking conflict about both the relevance of – and the demands on – knowledge. In the meantime, a vast number of studies examined the presence of non-academics in fields of (scientific) knowledge production (e.g. Brown et al. 2004; Shirk et al. 2012; Chilvers and Kearnes 2016; Lezaun, Marres, and Tironi 2017). This presence may manifest as forms of mobilising for knowledge-production, as in 'upstream engagement' (e.g. Rogers-Hayden and Pidgeon 2007) or 'citizen science' (e.g. Irwin 2002; Kimura and Kinchy 2019), to name two selected forms. Alternatively, this may happen in forms of self-mobilising for knowledge-production, as in conducting otherwise 'undone science' (Frickel et al. 2010). Brian Wynne termed this 'invited' and 'uninvited' (self-inviting) forms of participation, respectively (Wynne 2008). Similar dynamics can be shown with regard to erent levels (identity by practising an epistemic culture, identity by being a member of the world, identity by being an expert) – sometimes around strategic reasons or on grounds of nderstanding.

n up this first part related to identity and identity politics, we suggest a differentiation weakly tied identity politics and strongly tied identity politics. Weakly tied identity politics erised by the circumstance that the representational dominates the affectional aspect of ving activities. Strongly tied identity politics come into play when there are indivisible identity under attack. This is typically the case, if forms of 'embodied identity and otherness', entity and otherness' and 'located identity and otherness' are highly intertwined.

servation that the forms and aims of identity politics differ significantly between CSOs and brings us to the question of cooperation between these two groups in research projects. t-type format leads to increasingly formalised funding conditions. These are of significant e for the cooperation structure within participatory research. In the CONSIDER project, we fferentiating between two dimensions of interaction: the social interaction scheme on the versus the knowledge production scheme on the other. Doing so facilitates analysis of o achieve equilibrium between both dimensions in research processes involving co-work esearchers and CSOs (for the following, compare Böschen and Pfersdorf 2014). First, there *nteraction scheme*. In projects 'driven' by CSO involvement, CSOs interact intensively with sortium partners and are typically leaders of the consortium. At the other end of the a CSO's position is 'distant' from the core of the project, implying a minimum degree of action, e.g. as a cooperation partner of one consortium partner. Between these extremes ed' arrangements in which the level of social interaction among CSOs and other project largely equal. It is obvious that these interaction schemes only offer a snapshot of ted process of interaction. Secondly, the *knowledge production scheme* can be shown to ve dimension. In terms of knowledge production, CSOs can have either a 'focused' or native' role in a research project. Such a focused knowledge production scheme should be standard. CSOs thereby offer some selected piece of knowledge or are responsible for on tasks. This description of focused importance within knowledge production contrasts h that of a transformative knowledge production scheme. In such cases, CSOs have le impact on the core production of scientific knowledge. This core can be described nentary aspects of the production of knowledge, research agenda, methodology, empiri- and data collection and the interpretation of findings.

to imagine that cases involving a transformative impact on the production of scientific are more interesting with regard to the analysis of identity politics. In such cases CSOs e process that is otherwise entirely controlled by scientists, thereby challenging scientific d epistemic control. With regard to analysis of the three cases presented in the following has to ask how the cooperation is structured, whether identity politics is encouraged socio-epistemic tactics are used for constructing and maintaining social as well as ontrol in these hybrid forums of knowledge-production. Against this background the npirical analysis aims to shed light on the use of socio-epistemic tactics by the two actor wing both to conduct the project as well as to preserve their respective identities.

rmative Influence of CSOs: Three Selected Research Projects

al basis of this paper is an EU-funded research project: CONSIDER (2012–2015). The ysed whether – and if so how – CSOs are involved in research projects funded by the funding organisations. In summary, 33 case studies were conducted (for details of the sign, see Revel, Spruyt, and Soubiran 2013). Viewing projects from that perspective, estions arise concerning whether each of the six types of knowledge-production nanticipated finding was that all of the combinations of the social interaction scheme , CSO-balanced and CSO-driven) and the knowledge production scheme (focused

innovation processes, where it is closely linked to the idea of open innovation (e.g. von Hippel 2005). One highly important consideration, thereby, is that these participative activities typically go along with processes of citizens becoming 'formatted' in specific ways and towards selected role models, for example as 'bare citizens' (e.g. Braun and Schultz 2010) or 'users' (e.g. Hyysalo, Jensen, and Oudshoorn 2016).

These developments demand fresh scrutiny, especially (despite some relevant difficulties in realising participation) with regard to the selection of forms of knowledge seen as relevant (e.g. Bogner 2010), the often fragile and conflict-ridden interaction between different actors and actor groups (e.g. Klenk and Meeham 2017) or the fact that participatory measures are in great demand in the political sphere. Consequently, the participation of civil society actors is seen as a key resource for improving processes of research and innovation. Especially with the European Commission's guiding principles on Responsible Research and Innovation (RRI) and Open Innovation, we can observe a multifaceted set of strategies for mainstream participation in processes of knowledge production (e.g. Owen, Bessant, and Heintz 2013). The main idea behind these developments is not only to create legitimacy and acceptance for innovation, but also to improve innovations themselves. This is to be achieved by constituting 'co-responsibility' (Schomberg 2013) of civil society actors. This policy of mainstreaming 'epistemic inclusiveness' in research projects can be expected to create new constellations of social and epistemic control which, as starting hypotheses, call for new analytical tools for a better understanding.

The format of research projects offers a space to handle the expectations of research work, to organise research processes, to achieve selected outputs and to fix outcomes (Torka 2006, 63). While doing so, this framework restricts the principal openness of research, but potentially intensifies problems of coordination, especially if the actors involved are not exclusively from a research background. Thus, questions of cooperation become more challenging in these formalised project-types of cooperation. We argue that while projects with heterogeneous actors are driven by a dynamic to enable a balance between social and epistemic control, identity politics emerge. The issue of identity politics is highly relevant to the forms of participatory research. This perspective was introduced by Epstein (1996) and in the context of research on new social movements (Bernstein 2005, 53–55). However, the issue of identity and identity politics has not yet been systematically analysed in the literature on transdisciplinary research projects. This relates to the fact that long-term movements or overarching structural conditions of societies were typically seen as analytical starting points for investigations into identity politics (overview: Bernstein 2005).

This paper is divided into four sections. First, we will argue that forms of identity politics are related to forms of research projects. In doing so, we differentiate between weakly tied and strongly tied identity politics (Section 2). Secondly, we describe three cases analysed by the CONSIDER project (2012–2015). These cases are specifically selected to demonstrate the transformative impact of CSOs on knowledge production. It can be shown that their forms of social interaction differ considerably, from a CSO-distant mode to a CSO-driven mode (Section 3). Thirdly, the insights from the cases will be interpreted against the background of the named forms of identity politics (Section 4). Finally, some conclusions will be drawn from the conceptual considerations and empirical insights from these projects for the analysis, support and performance of transdisciplinary research (Section 5).

2. Identity and Identity Politics in Participatory Research Projects

The notion of identity is a concept used by a wide range of disciplines and which is highly contested with regard to political as well as theoretical-conceptual considerations (e.g. Brubaker and Cooper 2000; Schwartz, Luyckx, and Vignoles 2011). Nevertheless, the notion of identity is of increasing importance with respect to the break-up of entangled and therefore reciprocal stabilised relationships of agents, institutions, cultural classifications and political economy (e.g. Bernstein 2005). The question of identity gains its importance from the fact that identity is fundamentally situated at the edge of agency and politics (Hall 2000, 16). Identity comes into play in many cases, acting to roll out political agency by advocating for recognition of and respect for specific cultural or normative differences causing in many cases grievances or illegitimate economic inequalities. The claim of identity is thereby constitutive for allowing both to assemble a more or less homogeneous group that differs from other groups in some relevant respect, as well as for urging the necessity of political action.

Especially in the context of social movement research, these activities were seen as efforts to regain control over decisions and to expand the area of choice by seeking recognition for specific lifestyles or worldviews (e.g. Epstein 1998; Polletta and Jasper 2001); and in discourses on sexual orientation, it could be observed that these movements 'employed novel tactics and set out to challenge the very categories of identity that had previously motivated activism' (Bernstein 2005, 56). It is important to notice that identity cannot be seen as a fixed category: On the contrary, it is a concept relating to continuous processes of identification with broader communities, practices or institutions for self-positioning and self-understanding. This has consequences for politics and vice versa. While governing by and through identity, identity is continuously in a state of flux (Lawler 2014).

With regard to the question of identity politics within research projects, four points require discussion. First, there is the question of how we can understand and conceptualise identity here. This discourse is multifaceted, such that we have to narrow down the analysis to some of the most important aspects. Secondly, we must ask how we can actually understand CSOs and their importance for identity politics in the public sphere. This issue underlies the often complicated or contested relations to the represented communities, and needs to be disentangled analytically. Thirdly, we must view science and scientists as key actors in research projects. In which senses are identity and identity politics relevant for them? Fourthly, it is important to look specifically at research projects concerning cooperation in knowledge production processes. We do so by offering a typology specifically elaborated for understanding the cooperation of CSOs with scientists in research projects.

One often-cited definition claims that identity is 'that part of an individual's self-concept which derives from his knowledge of his membership of a social group together with the value and emotional significance attached to that membership' (Tajfel 1978, 63). This definition refers to social identity theory. Nevertheless, there are good reasons to use the term 'collective identity' instead (Ashmore, Deaux, and McLaughlin-Volpe 2004, 81). Identity is a highly relevant aspect of any community or social movement, but it is constituted differently and has to be described multi-dimensionally (e.g. Breakwell 1986; Ashmore, Deaux, and McLaughlin-Volpe 2004, 83). With regard to environmental psychology, the category of 'place' must also be taken into account as a strong variable (e.g. Lewicka 2011; Clarke, Murphy, and Lorenzoni 2018). Building collective identity is a two-sided process as it addresses 'we' as well as 'them', or 'intern' as well as 'extern' classifications (for an overview see Wetherell 2010). These categorisations allow the construction of otherness by 'we' and 'them', and the performance of identity politics. Firstly, there is the aspect of being specific, by having a shared collective identity. Secondly, and with regard to the external reference: Identity is a claim for saying what is different and 'other' in comparison to other social groups. Both aspects together enforce a variety of different goals, such as the protection of their own social sphere against other 'worlds', the fight for acceptance of a hitherto stigmatised identity or rearranging the public perception of selected social groups by challenging the classifications used.

However, these politics for separating social spheres by claims of identity are not only fuelled by those claiming a specific identity. Often, there is also a dynamic of 'othering' by social groups that are not identified with the identity under consideration. Seen that way, identity politics is two-sided, because processes of self-positioning and other-positioning are aligned together. The literature discusses three main forms of 'otherness' and 'othering'. These are, firstly, forms of 'embodied identity and otherness', typically related to all forms of being other by specific characteristics that can be related to the body, such as being male or female, or sexual orientation (e.g. with respect to health movements: Brown et al. 2004). Secondly, there are forms of 'cultural identity and otherness'.

These are often related to different forms of religious orientation, as religi
the core set of orientation patterns in a cultural sphere, but are not restrict
Cooper 2000). Thirdly, there are forms of 'located identity and otherness'. T
at concrete places through which they connect their feeling of identity
Lorenzoni 2018). Easy to imagine, these different forms of 'identity and oth
other and might overlap (e.g. Blume 2010).

What about CSOs and their role and relation to identity and therefore i
not necessarily the same as movements or communities. On the conti
a specific representational relationship to one single or a couple of comi
a widely used definition by the European Commission (2008, 6), '[...] are c
entity that is non-governmental, not-for-profit, not representing commerci
a common purpose in the public interest'. Thus, the term CSOs subsumes
social groups. To some degree, they can be placed between activist move
They act like movements in that they identify problems that require societa
identities for specific groups represented by them and defining who is
problem (e.g. Barbot 2006; Rabeharisoa, Moreira, and Akrich 2014).

Identity politics are performed by using selected tactics. Typical epister
selected features of knowledge production, e.g. by offering specific kn
focusing on otherwise ignored questions (Frickel et al. 2010), promoting sel
(Callon and Rabeharisoa 2008) or contesting methodological presuppositio
tactics aim to control epistemic claims within the cooperation by expandin
form of cooperation and vice versa. Forms of identity politics will depe
consideration (as these are related to both variants of epistemic involvem
forms of identity and otherness), the form in which CSOs are included within t
the relation of CSOs to communities. CSOs might have a claim to represent a
in the political arena, claiming public interest while promoting specific inter
may represent a group with special interests, promoting these special ir
acknowledged and protected by society.

Questions of identity and otherness must also be addressed to science.
one can state: 'Identity is an assertive element for the practitioners of a disci
concept of affiliation developed through years of immersion in a cognitiv
specific concepts, a specific vocabulary, and grounded in a family of phenor
ments, theories; all reinforced by "habitus" and the professional association
(Marcovich and Shinn 2011, 588) This means that scientists are socialised and t
which by immersion are experientially lived as identity. Identity is not to b
'epistemic identity' given by the epistemic culture and skills in which indivic
this culture are trained. The influence of an 'epistemic identity' can be show
interdisciplinary cooperation between researches practising different epistem
flicts emerging in such settings are closely related to the fact that these conve
to varying degrees by safeguarding one's own 'epistemic identity' in the face o
by the presence of other ways of 'epistemic world-making'.

However, there is a powerful presupposition of 'being a scientist' (sharing
a member of the scholarly world). There are shared, value-based identities,
autonomy which is important for all sciences. Autonomy describes a basic
allowing scientists to work detached from specific societal affordances. Not su
tant socio-epistemic tactic employed by scientists in hybrid forums is to
knowledge-production. For example, 'citizen science' activities invite lay-peop
time protecting scientists from being questioned in their role (Kimura and Kin
mirrored in one important generalised classification: that between 'experts' o
'lay-people' on the other, resulting in the claim of the so-called 'deficit model'
2015). Obviously, identity and identity politics for and by scientists can be im

these di
scholar
self-mis
To s
betwee
is chara
identity
aspects
'cultura
The
scientis
The pro
import
sugges
one ha
attemp
betwee
is a *soc*
other
spectru
social
are 'ba
partne
a mult
be a
a 'tran
seen a
dissem
sharp
a rem
by the
cal de
It i
know
influe
autho
sectio
and v
episte
follow
group

3. T

The e
proje
EC ar
empi
differ
occu
(CSO

impact, transformative impact) could be observed (Böschen and Pfersdorf 2014, 42). The following section analyses three exemplary cases offering insights into the form and intensity of identity politics in relation to projects where CSOs have a transformative impact on knowledge production. The three projects selected are: (a) a community-based research project (CSO-driven, transformative impact on knowledge production; case KIT A)[1], (b) a cooperative–inclusive research project (CSO-balanced, transformative impact on knowledge production; case KIT F) and (c) a peripheral-inclusive research project (CSO-distant, transformative impact on knowledge production; case KIT J). One important factor should be noted: As the researchers gained access to some of the projects only on condition of anonymity, the projects are necessarily discussed in quite an abstract manner in order to avoid potentially disclosing their identities. Narratives are typically enriched by these tastes of concreteness; consequently, such richness is unfortunately lacking from the accounts presented here.

3.1. CASE A: Community-based Research Project following an Environmental Disaster

This project was set up in the context of a severe crisis involving widespread environmental contamination following an industrial disaster. Subsequently, the state administration saw the need for such a project to outline the impacts of chemical discharges on the environment and possible strategies to cope with the associated damage. Different local communities lived within this environment, and were the main beneficiaries of the research project. The relevance of the project was confirmed by an exploratory phase that revealed contamination of soil and fauna in the area where the communities live. This was relevant because the food sector provides one pillar of the communities' economic basis. As one of the communities had already established trusted cooperation with an environmental CSO over a period of several years, the CSO initiated the process through an invitation to a research institute well known by them. Demonstrating their willingness to cooperate further, additional local communities and CSOs came on board. The project itself aimed to analyse the distributions of environmental pollutants at sites and objects that were identified as highly relevant for the local communities, and to assess their impacts on human health and economic conditions. Fishery was one main source of income for these local communities. Within this project the CSOs and participating communities had a strong influence on the research process, as it covered various aspects within the process of knowledge production (including the definition of the research agenda, methodologies as well as data capture and interpretation). Against this background, a complex landscape of organisational as well as communicative measures was established to structure the co-work between these groups on the one hand and the researchers on the other. Nevertheless, the socio-epistemic tactics differed between the groups.

With regard to local communities and CSOs: Firstly, it was important to identify the hazards and damage within the study area that were relevant to the local communities and, if possible, to remediate this damage. In this respect, analysing and controlling the effects on natural environment with regard to selected endpoints within the ecosystem were key. The local communities followed a tactic of endpoint-oriented research, which forces the researcher towards difficult decisions (see below). Secondly, a decisive act was to exclude industry and their partners from the project. Therefore, industry was not part of the project, either as a partner or a subcontractor. This was seen as a precondition, as industry caused this severe environmental damage but yet was unable to offer the communities acceptable forms of compensation. Moreover, it was assumed that the knowledge production process would have been negatively influenced by the inclusion of industry interests in the project. Thus, for the communities, one key socio-epistemic tactic was to radically exclude industry and their knowledge from the project (including preventing the researchers from contacting industrial experts).

The researchers, confronted with this tactic of endpoint-oriented research faced a problem, as this represented a call for exceptions from methodologically sound standards:

> Well, there is a limitation in seeking the truth and that's why I'm sending [my PhD-student] to go and find out {laughs} what everybody else is doing. What I'm saying is we are not presenting our data. But we are trying to find out what other people are finding. Is the data we've got – from the places that she's got on her map – different from what other people have found? That is the reason why we want to do that. ... She's a spy. (KIT A_INT 1)

The underlying epistemic problem was that the researchers had to modify their data collection routines and therefore faced the methodological problem of determining whether or not the collected data were suitable. Moreover, these boundary conditions meant that it was not possible to successfully work towards completing a PhD thesis within the framework of the project. Therefore, the key socio-epistemic tactic consisted of reframing their work as providing services. This was astonishing with regard to the epistemological innovation that would have been possible had the researchers followed the non-routinised forms of knowledge production. However, this was seen as too risky. Thus, there was an interesting constellation, wherein the CSOs saw the necessity of conducting risky research whereas the researchers shrank back from this. Importantly, the scientists encountered conflicts for preserving their epistemic identity as well as authority through the socio-epistemic tactic of self-distancing from their work on the project.

3.2. CASE B: Cooperative-inclusive Research Project for Promoting Sustainability Innovations

This project was funded as a stream of research for the benefits of including CSOs partners. Three groups of organisations played major roles in this research project: local communities, CSOs and research institutions. The basic idea of the project was to understand whether and how selected communities attract themselves to specific sustainable social innovations, to model the social and environmental conditions that enable such innovations and finally to transfer these innovations to other communities. The geographical focus of the project was Latin America. There, the researchers identified four different local communities that implemented social innovations that were sustainable and improved living conditions in their respective communities. One case in point concerned payments for ecosystem services. For example, a community changed its method of producing bricks and thereby protected the environment. As the products were ecologically valuable, the community received a payment. Nevertheless, the researchers lacked a robust connection to the communities. Thus, the inclusion of CSOs that had established such connections through years of cooperation was decisive not only with regard to the funding scheme but also to functional necessities, and was greatly appreciated by the scientists involved in the project.

The CSO project partners possessed both local knowledge and connections of trust to the different local communities. Some were experienced in working with these communities and all were specialists in issues concerning sustainability. Moreover, they had local knowledge at their disposal. This was necessary for conducting the innovations for sustainability. Therefore, the project as such would not have been conducted without the CSOs, but they also profited from it. The project supported them in extending their role in the region. The CSOs strengthened their connection to the communities and also gained societal recognition through their growing scientific reputations. Being involved in agenda setting and data collection and being responsible for field trips conducted by the research team, the CSOs gained vast experience that was highly valuable for their work. Moreover, one interviewee from a community emphasised that it was important for communities as well as CSOs that their work was now well documented. Seen that way, the CSOs used the well-known socio-epistemic tactics of acquiring scientific knowledge and enforcing their role as representatives of selected communities (e.g. Epstein 2008). Interestingly, the researchers themselves saw the fundamental importance and need for CSO involvement in making the project a success – not simply for meeting the requirements of the funding agency, which called for CSO involvement as a prerequisite for funding. The scientists conceded that:

> You need to adapt science to the needs of those who are the end-users, for instance in the case of technologies, or where the actual problem has emerged from. You need to make the connection which might be missing in the realm of pure science. Taking this aim seriously, we involve CSOs in different stages of the project – not only in the end for dissemination but also in the beginning for agenda setting (KIT F_INT 2).

Hereby, one decisive success factor was the project's chosen framework, which followed a methodology of action research:

> We are following this approach of action research and we are working together with the local stakeholders. This is decisive for shortening the procedure. Well, the project is only running a short time and thus there is a need to start as soon as possible. You need partners being accepted and you trust in them. (KIT F_INT 1)

The framework of action research worked well and the scientists treat the CSOs more or less as an equal in knowledge production. The researchers profited from the cooperative arrangement because the CSOs facilitated data collection and data analysis. Against this background, they and the scientists published scientific findings together. It was decisive for the success of the project that both researchers and CSOs were oriented towards adaptation of the concrete option for innovation in the respective environments of the communities. This means that they followed the idea of creating a social learning field to contextualise, disseminate and mainstream sustainability innovations. Proceeding in that manner, all the different actors had both a relatively broad space for gaining knowledge as well as a clear orientation to conduct the research. One can claim this as a socio-epistemic tactic of shared experimental learning. This project was spurred on by the fact that researchers as well as CSOs were convinced of the importance of this innovation and the need to contextualise it. Consequently, both groups easily collaborated in trialling the respective innovations within different local contexts.

3.3. CASE C: Peripheral-dominant Research Project: Health Technology Innovations

This research project was funded by the German Ministry of Education and Research (BMBF). By the end of the research project, the participants expected to develop some technical applications for disabled people as the most obvious outputs. Moreover, the researchers expected to gain insights about processes of ageing and to learn how the bodies of elderly people function. Apart from these results, the project also aimed to produce methodological knowledge about the possibilities of end-user integration into a scientific and technological research project. In contrast to typical research proposals, the specific content-related focus of the project was not described in advance. The disciplinary knowledge, competences and abilities of the scientists and company workers involved in the project limited the range of possible outcomes. Two other technical applications resulted from the participatory research process, and were previously unknown. The interdisciplinary research group involved scientists from research institutes as well as employees from a hospital and the business sector. The group co-developed common ideas for products and applications together with a selected group of disabled people who represented an exemplary group of end-users. Their knowledge was the foundation of the R&D-process in this project. Therefore, they were involved in the knowledge production by obtaining and interpreting the data and having authorship of articles and patents. There was a stimulating convergence of research interests and the orientation towards marketable products. These issues were scientifically challenging, and solving them held the promise of delivering market-relevant innovations. Primarily, representatives from the CSOs representing the disabled users took part as affected persons. With greater project progress, these representatives felt it could also be helpful to spread knowledge about the new applications to the interested community.

Most challenging was the design of the cooperative arrangements between the scientists and the disabled groups, with respect to the CSOs. The two main approaches for including external actors within the research process consisted of online social media platforms and workshops involving disabled end-users. This imposed a distance from the core team involved in the research and development (R&D) process. This construction was instructive, as the CSOs operated both inside and outside the project at the same time. As the CSOs were simultaneously in a socially distant position, conflicts emerged and we found several strategies for conflict resolution, e.g. mutual agreement or consent regarding general decisions on the project or specific questions of product development were to be answered by CSOs. The important socio-epistemic tactic for the researchers was to parcel out the research process in a clear set of steps with specific interaction options. This can be seen as a tactic of selected affiliation. For example, cooperation with the technical developers was also placed at one specific step distant from the development. Moreover, there was an interesting convergence of prototyping and scientific knowledge-production. This can be seen as a tactic of strategic coupling as it allowed scientific knowledge-production (as latent interest) to be conducted under the promise of creating new treatment opportunities (manifest interest). To put it in slightly exaggerated terms, this can be interpreted as epistemic camouflage.

The cooperation between researchers and the disabled groups could have been established through the work of an intermediary that had worked for more than ten years with both the CSOs and disabled users. This intermediary was decisive in making this project possible, since the disabled actors were uneasy about such collaborations, having previously had many negative experiences. The CSOs and the respective disabled actors wanted to be part of the social network and to be recognised as individuals by the researchers. They therefore followed the socio-epistemic tactic of building a problem-solving community. This collided with the call for autarky on the side of the researchers. Nevertheless, the project was successful as the coordinators not only put an intermediary into action but conducted the project using a series of platforms for interaction following a tactic of selected affiliation. As a result, the latent potentials for protest by the CSOs could be canalised and the researchers' desire for demarcation could be served.

4. Discussion: Epistemic Conflict and Identity Politics

The following descriptions should be viewed as portrayals of finished processes. Analysis of the three cases highlights three insights. First, the projects differ with regard to the socio-epistemic tactics used by the groups involved. These mixtures of tactics are closely related to the forms of the projects studied here. The use of such tactics corresponds to the epistemic conflicts upcoming in the projects. Thus, they serve to control one's own social and epistemic territory against possible unwanted or negative impacts associated with the cooperation partners. Second, the differentiation of the two analytical dimensions of 'knowledge production scheme' and 'social interaction scheme' and their interplay shows in the three cases that each concrete form of epistemic and social control represents the outcome of an equilibrium between social as well as epistemic interactions. Third, not all the cases show that these kinds of projects are necessarily highly conflict-driven processes of interaction between researchers and CSOs in the course of the mostly problem-centred knowledge production. Although problem-centred knowledge production seems to introduce a relevant tension into these actions, this is the tension between researchers, who must explore new ways of knowledge production for problem-solving, and CSOs that often focus on very specific aspects of the whole problem in seeking to ensure that their particular problem is solved. The pressures of identity politics differ, and were not necessarily high or else could be reduced. Although a significant variety of socio-epistemic tactics was to be observed, their interplay and the resulting cooperative structure seemed to be successful, in the sense that the room for cooperation was structured without closing it down. Nevertheless, the outcomes were different, as in cases A and B we observed a shared epistemic control in contrast to case C.

In case A the pressure for scientists was decreased, as the problem-solving approach as well as the communities' responsibility for the project organisation offered space for the tactic of self-distancing. In doing so, they avoided conflict with their own scientific community, but were not able to test whether the situation offered an option for new or innovative ways of producing knowledge. For the communities, this was impossible, as the environmental catastrophe within the study area directly impacted their natural habitat and ways of life. Thus, identity-saving measures were key for them. Therefore, this project only succeeded because the communities' strongly tied identity politics was accompanied by a weakly tied identity among the scientists.

Case B was unproblematic with regard to identity politics as all of the actors followed the socio-epistemic tactic of shared experimental learning, thereby leaving plenty room for each actor and allowing far-reaching cooperation within the project as the division of labour was easily defined and allowed CSOs to work scientifically as the researchers. The CSOs had the trust of the communities and could integrate local knowledge; the researcher had the role of generalising the outcomes of the different local experiences with regard to the generalisability of the innovation under consideration. Importantly, neither group (i.e. neither researchers nor CSOs) experienced severe conflicts with regard to maintaining identity. This means that both main actor groups displayed weakly tied identity politics.

In contrast to cases A and B, case C appears different as there was a need for strongly tied identity politics among the relevant actor groups. Anticipating this, the researchers adopted two tactics: strategic coupling of research goals, and intermediation, thereby allowing them to hold epistemic control. Nevertheless, the entire research process was difficult, especially for the CSOs and the disability actors that they represented. They had to agree to share their knowledge, although they had to fight against any form of epistemic exploitation of their specific situation and body. Their goal was mainly oriented towards protecting the group against any kind of infringement while sharing knowledge. The main socio-epistemic tactic was to enforce a problem-oriented approach to knowledge production by building a problem-solving community. This was in strong tension with the tactic of selective affiliation, as they together represent two different identity concepts: one of 'connectedness' (selective affiliation tactic) and 'grouping' (a tactic building a problem-solving community) (for the differences between connectedness versus grouping, see Brubaker/Cooper 2000:19f.). This resulted in a seemingly unsolvable dichotomy of identity politics, which was ultimately addressed by implementing a strong project governance regime, protecting the researcher's epistemic control by structured social control.

5. Conclusions

Looking from these findings back to the starting point of our journey, it is apparent that scientific identity politics was formerly a comparatively easy task. What are the overarching lessons to be drawn from this analysis? Firstly, there are highly productive approaches for including CSOs in research, which also facilitate transformative forms of knowledge production. Nevertheless, in most cases, these are coordination-intense processes, as the forms of social as well as epistemic control are bargained throughout the entire project. Both researchers and scientists have their own tactics for offering themselves the option of participation. In any case, such collaboration is a risky endeavour for both groups, as it might challenge their routines and premises of action. Secondly, these tactics are closely related to the need for preserving the identities of the communities to which the respective actors feel attached. This provokes tactics of identity politics, pushing needs for an explicit governance within the project team. These tactics are selected in order to realise forms of epistemic and social control. Thirdly, as forms of co-research expand, questions of who then owns the process – and what are the consequences for the respective identities of the communities – become urgent. Thus, the politics of scientific identity is becoming an undeniable task and, moreover, a demanding one. This is not difficult to understand, as these co-research actions offer a hybrid forum between different communities and their understandings of what is (and what is not) relevant knowledge for problem-solving. However, as the three case studies show, each of the projects was ultimately successful and each delivered meaningful results. This was attributed to different reasons in each case.

Relating this back to established and upcoming research traditions, these insights are of great importance. Looking at discourses with a strong tradition of designing as well as analysing the interaction between researchers and civil society actors, namely those of 'action research' (locus classicus: Lewin 1952; e.g. Kemmis and McTaggert 2005) and of 'transdisciplinary research' (e.g. Pohl, Krütli, and Stauffacher 2017; Felt et al. 2016), the upcoming conflicts within the research processes were mostly analysed as resulting from diverging interests or an unequal power structure. In the meantime, interesting new alliances and couplings between these strands as 'living labs' or 'real-world laboratories' (e.g. Schäpke et al. 2018) take place in the research landscape. These forms of real-world experimentation are gaining importance by the fact that there are concrete, real-world questions to be solved which require the inclusion of citizens or their representatives (CSOs) as a more or less pivotal prerequisite. Moreover, the traditions mentioned above have opened up a sophisticated and heterogeneous field of practical experimentation, but also theoretical consolidation. Thereby, some of the basic assumptions and foci of analysis were used without specifically reflecting their adequacy under the new boundary conditions.

Thus, as these developments indicate a reorientation of knowledge production within processes of societal problem-solving, the analysis of identity politics as presented in this article seems to be a building block for uncovering the ordering of social as well as epistemic control in such processes. One important insight from the literature so far was that socio-epistemic tactics were more part of the problem than of the solution. Surprisingly, the findings of the present article offer insights into a situational fit. This was the case if at least one of the key groups had the opportunity to follow a weakly tied identity politics. Moreover, one might argue that the 'project' format of the case studies calmed the conflicts and restricted the range of tactics used. With regard to the emergence of more open formats of research such as those addressing real-world experiments, one can presume that the formal constitution of such processes might help to moderate conflicts between the socio-epistemic tactics employed by what are inherently heterogeneous collections of actors.

Note

1. The cases are named first after the institution responsible for conducting the empirical research, and then a letter indicating the internal number of the case.

Disclosure statement

No potential conflict of interest was reported by the authors.

Funding

We thank the European Commission for funding. CONSIDER was funded as part of the EU Framework Programme FP7-Science-in-Society-2011-1 under the grant agreement number: [288928].

ORCID

Stefan Böschen http://orcid.org/0000-0003-0519-5030
Bernd Carsten Stahl http://orcid.org/0000-0002-4058-4456

References

Ashmore, R. D., K. Deaux, and T. McLaughlin-Volpe. 2004. "An Organizing Framework for Collective Identity: Articulation and Significance of Multidimensionality." *Psychological Bulletin* 130 (1): 80–114. doi:10.1037/0033-2909.130.1.80.

Barbot, J. 2006. "How to Build an 'Active' Patient? The Work of AIDS Associations in France." *Social Science & Medicine* 62: 538–551. doi:10.1016/j.socscimed.2005.06.025.

Bernstein, M. 2005. "Identity Politics." *Annual Review of Sociology* 31: 47–74. doi:10.1146/annurev.soc.29.010202.100054.

Blume, S. 2010. *The Artificial Ear: Cochlear Implants and the Culture of Deafness*. New Brunswick, NJ: Rutgers University Press.

Bogner, A. 2010. "Let's Disagree! Talking Ethics in Technology Controversies." *Science, Technology & Innovation Studies* 6 (2): 183–201.

Böschen, S., and S. Pfersdorf. 2014. "Model of CSO Participation in Research Governance." *CONSIDER: Deliverable D3.3*. Karlsruhe (Ms).

Braun, K., and S. Schultz. 2010. "A Certain Amount of Engineering Involved. Constructing the Public in Participatory Governance Arrangements." *Public Understanding of Science* 19 (4): 403–419. doi:10.1177/0963662509347814.

Breakwell, G. M. 1986. *Coping with Threatened Identity*. London: Methuen.

Brown, P., S. Zavestoski, S. McCormick, B. Mayer, R. Morello-Frosch, and R. G. Atman. 2004. "Emobdied Health Movements: New Approaches to Social Movements in Heath." *Sociology of Health & Illness* 26 (1): 50–80. doi:10.1111/j.1467-9566.2004.00378.x.

Brubaker, R., and R. Cooper. 2000. "Beyond 'Identity'." *Theory and Society* 29: 1–47. doi:10.1023/A:1007068714468.

Bucchi, M. 2015. "Changing Contexts for Science and Society Interaction: From Deficit to Dialogue, from Dialogue to Participation – And Beyond?" In *The Public Shaping of Medical Research*, edited by P. Wehling, W. Viehöver, and S. Koenen, 211–225. London: Routledge.

Callon, M., P. Lascoumes, and Y. Barthe. 2009. *Acting in an Uncertain World. An Essay on Technical Democracy*. Cambridge, MA: MIT-Press.

Callon, M., and V. Rabeharisoa. 2008. "The Growing Engagement of Emergent Concerned Groups in Political and Economic Life." *Science, Technology & Human Values* 33 (2): 230–261. doi:10.1177/0162243907311264.

Chilvers, J., and M. Kearnes, eds. 2016. *Remaking Participation: Science, Environment and Emergent Publics*. London: Routledge.

Clarke, D., C. Murphy, and I. Lorenzoni. 2018. "Place Attachment, Disruption and Transformative Adaptation." *Journal of Environmental Psychology* 55: 81–89. doi:10.1016/j.jenvp.2017.12.006.

Epstein, S. 1996. *Impure Science: AIDS, Activism and the Politics of Knowledge*. Berkeley: University of California Press.

Epstein, S. 1998. "Gay and Lesbian Movements in the United States: Dilemmas of Identity, Diversity, and Political Strategy." In *The Global Emergence of Gay and Lesbian Politics: National Imprints of a Worldwide Movement*, edited by B. D. Adam, D. W. Duyvendak, and A. Krouwel, 30–90. Philadelphia, PA: Temple University Press.

Epstein, S. 2008. "Patient Groups and Health Movements." In *Handbook of Science and Technology Studies*, edited by O. Hackett, M. Amsterdama, M. Lynch, and J. Wajcman, 499–540. Cambridge, MA: MIT Press.

European Commission. 2008. *Commission Recommendation on 07/ 02/2008on a Code of Conduct for Responsible Nanosciences and Nanotechnology (No. C(2008) 424 Final)*. Brussels: European Commission.

Felt, U., J. Igelsböck, A. Schikowitz, and T. Völker. 2016. "Transdisciplinary Sustainability Research in Practice: Between Imaginaries of Collective Experimentation and Entrenched Academic Value Orders." *Science, Technology, & Human Values* 41 (4): 732–761. doi:10.1177/0162243915626989.

Frickel, S., S. Gibbon, J. Howard, J. Kempner, G. Ottinger, and D. Hess. 2010. "Undone Science: Charting Social Movement and Civil Society Challenge to Research Agenda Setting." *Science, Technology & Human Values* 35 (4): 444–473. doi:10.1177/0162243909345836.

Hall, S. 2000. "Who Needs Identity?" In *Identity: A Reader*, edited by P. du Gay, J. Evans and P. Redman, 15–30. London: Sage.

Hyysalo, S., T. E. Jensen, and N. Oudshoorn, eds. 2016. *The New Production of Users. Changing. Innovation Collectives and Involvement Strategies*. London: Routledge.

Irwin, A. 2002. *Citizen Science: A Study of People, Expertise and Sustainable Development*. London: Routledge.

Kemmis, S., and R. McTaggert. 2005. "Participatory Action Research: Communicative Action and the Public Sphere." In *Handbook of Qualitative Research (3rd Edition)*, edited by N. Denzin and Y. Lincoln, 559–603. Thousand Oaks: Sage.

Kimura, A., and A. Kinchy, eds. 2019. *Science by the People. Participation, Power, and the Politics of Environmental Knowledge*. New Brunswick, NJ: Rutgers University Press.

Klenk, N. L., and K. Meeham. 2017. "Transdisciplinary Sustainability Research beyond Engagement Models: Toward Adventures in Relevance." *Environmental Science and Policy* 78: 27–35. doi:10.1016/j.envsci.2017.09.006.

Lawler, S. 2014. *Identity. Sociological Perspectives*. Cambridge: Polity Press.

Lewicka, M. 2011. "Place Attachment: How Far Have We Come in the Last 40 Years?." *Journal of Environmental Psychology* 31: 207–230. doi:10.1016/j.jenvp.2010.10.001.

Lewin, K. 1952. "Group Decision and Social Change." In *Readings in Social Psychology*, edited by T. M. Newcomb and E. E. Hartley, 459–473. New York: Holt.

Lezaun, J., N. Marres, and M. Tironi. 2017. "Experiments in Participation." In *The Handbook of Science and Technology Studies*. 4th ed., edited by U. Felt, et al., 195–221. Cambridge, MA: MIT-Press.

Marcovich, A., and T. Shinn. 2011. "Where Is Disciplinarity Going? Meeting on the Borderland." *Social Science Information* 50 (34): 582–606. doi:10.1177/0539018411411036.

Owen, R., J. Bessant, and M. Heintz, eds. 2013. *Responsible Innovation*. London: John Wiley.

Pohl, C., P. Krütli, and M. Stauffacher. 2017. "Ten Reflective Steps for Rendering Research Societally Relevant." *GAIA* 26 (1): 43–51. doi:10.14512/gaia.26.1.10.

Pol, J. 2014. "Knowing Patients: Turning Patient Knowledge into Science." *Science, Technology, & Human Values* 39 (1): 73–97. doi:10.1177/0162243913504306.

Polletta, F., and J. M. Jasper. 2001. "Collective Identity and Social Movements." *Annual Review of Sociology* 27: 283–305. doi:10.1146/annurev.soc.27.1.283.

Rabeharisoa, V., T. Moreira, and M. Akrich. 2014. "Evidence-based Activism: Patients' Organisations, Users' and Activist's Groups in Knowledge Societies." *BioSocieties* 9 (2): 111–128. doi:10.1057/biosoc.2014.2.

Revel, M., E. Spruyt, and T. Soubiran. 2013. "FP7 Survey Report: Consider." *Deliverable 2.2*. Lille (Ms).

Rogers-Hayden, T., and N. Pidgeon. 2007. "Moving Engagement 'Upstream'? Nanotechnologies and the Royal Society and Royal Academy of Engineering's Inquiry." *Public Understanding of Science* 16 (3): 345–364. doi:10.1177/0963662506076141.

Schäpke, N., M. Bergmann, F. Stelzer, and D. Lang. 2018. "Labs in the Real World: Advancing Transdisciplinary Research and Sustainability Transformation. Mapping the Field and Emerging Lines of Inquiry." *GAIA* 27/S1: 8–11. doi:10.14512/gaia.27.S1.4.

Schomberg, R. 2013. "A Vision of Responsible Innovation." In *Responsible Innovation*, edited by R. Owen, J. Bessant, and M. Heintz, 51–74. London: John Wiley.

Schwartz, S. J., K. Luyckx, and V. L. Vignoles, eds. 2011. *Handbook of Identity Theory and Research. Vol. 1: Structures and Processes*. Berlin: Springer.

Shirk, J., H. L. Ballard, C. C. Wilderman, T. Phillips, A. Wiggins, R. Jordan, E. McCallie et al. 2012. "Public Participation in Scientific Research: A Framework for Deliberative Design." *Ecology & Society* 17 (2): 29. DOI:10.5751/ES-04705-170229.

Tajfel, H. 1978. "Social Categorization, Social Identity, and Social Comparison." In *Differentiation between Social Groups: Studies in the Social Psychology of Intergroup Relations*, edited by H. Tajfel, 61–76. London: Academic Press.

Torka, M. 2006. "Die Projektförmigkeit der Forschung." *Die Hochschule: Journal für Wissenschaft und Bildung* 15 (1): 63–83.

von Hippel, E. 2005. *Democratizing Innovation*. Cambridge, MA: MIT Press.

Wetherell, M. 2010. "The Field of Identity Studies." In *The Sage Handbook of Identities*, edited by M. Wetherell and C. T. Mohanty, 4–26. London: Sage.

Wynne, B. 2008. "Public Participation in Science and Technology: Performing and Obscuring a Political-Conceptual Category Mistake." *East Asian International STS Journal* 1 (1): 1–13.

The Phase Zero: Why Collaborative Research Is Not Co-Designed but Scripted

Jeremias Herberg

ABSTRACT
The very beginning of collaborative research endeavors often lies in politically difficult and practically challenging entanglements. The purpose of this paper is to empirically capture and theoretically conceptualize these entanglements. I trace the power-driven prefiguration of my own role in a transdisciplinary project and argue that the early moments (the 'phase zero') of collaborative research are entwined with a tacit, tactical, and relational form of control. In a process that I call 'scripting control,' actors seek to co-determine what a project may become, without being able to forecast or backcast a pathway to get there. Collaborators mutually launch counter-scripts and tacitly shape the possibility space that constrains or enables subsequent interactions. My own transdisciplinary involvement illustrates, however, that counter-scripts proposed by latecomers can fail if the project has passed the phase zero. This argument extends the current use of scripts in Science and Technology Studies to also involve temporal power dynamics. Moreover, in sustainability studies, my argument contributes to a growing critique against the imaginary of co-design, which promotes a managerial idea of ordering collaborative processes in a socially and epistemically inclusive way.

The initiating moments of collaborative research are often characterized by a practical quest for collaborators to preconfigure power relations. This article is the product of practical reflection on this very process. I reconstruct the genealogy of my own position as a transdisciplinary researcher at an emerging educational organization and propose a concept of control to discern, not only empirically but conceptually, how power relations are molded early on in collaborative research. The distinction between phases and the transition between different forms of control implies methodological, analytical and even moral difficulties. I tackle those based on the popular concept of 'scripts' that originated within Science and Technology Studies (STS). Scripts appear as a given, when for instance encountering a door that facilitates access to a building. Yet, they can be traced back to a much more volatile process of designing. This process, which in transdisciplinary sustainability studies (TSS) is affirmatively described through the notion of 'co-design' (e.g. Moser 2016), essentially implies a form of control that has not been explicated. Through a relational and highly dynamic process of scripting, I argue, the control regime of collaborative research is established.

In section one, I discuss the dominant idea of control in sustainability studies, making room for an alternative understanding of control that I present in the second section based on a spatiotemporal understanding of 'scripts' (Akrich and Latour 1992). In section three, I show in a case study how, with emphasis on mutually ascribed scripts, a collaborative project emerges. Section four discusses the difference between phase zero and phase one in regard to my own quest for counter-scripts. Section five discusses the normative implications based on Paul Virilio's notion of 'dromocracy' (Virilio [1977] 2006).

1. The Imaginary of Design Control in Sustainability Studies

Social theorist Barbara Adam wrote in her seminal work about time that 'the quest for control is to a large extent about obtaining dominion over time for economic gain and social advantage' (Adam 2004, 123). In a book together with Chris Groves, she coins the phrase 'present futures' to describe this ethos of control (Adam and Groves 2007, 161–187): Time is classified and measured in order to fit the linear coupling of problems and solutions, which characterizes the dominant practices of business and policy making. One prominent approach, which emerged as a sophisticated continuation of linear control, goes back to systems analysis in the United States in the 1960s. In the face of thorny problems such as military planning or the allocation of welfare funds, the Lyndon B. Johnson administration and the RAND Corporation endorsed an adaptive solution-finding methodology, identifying processes and developing corresponding techniques such as cost–benefit analysis and program budgeting. According to political scientist Aaron Wildavsky, these techniques revolved around an iterative principle of design: 'New objects are defined, new assumptions made, new models constructed, until a creative amalgam appears that hopefully defines a second-best solution, one that is better than others even if not optimal in any sense' (Wildavsky 1966, 301). Systems analysts, most famously US Secretary of Defense Robert McNamara, proactively endorsed this dynamic approximation of means and ends during the course of the Vietnam War and even in social policy making (Jardini 2000). However, as the sociologist Ida Hoos criticized early on, this approach undermined ethical decision making and democratic leadership. The failure to control for unintended consequences, which was painfully revealed and publicly scrutinized during the Vietnam War, only increased the centralization of control and promoted a sweeping economization of military strategy and social policy (Hoos 1972). Hence, even the most reflexive methods of control can be used to evade questions of responsibility and sustainability. Adam and Groves therefore advocate for the alternative concept of 'future presents' [...] which 'positions us with reference to potential impacts of present actions on future generations who have to cope with the consequences of our inventions and interventions.' (Adam and Groves 2007, 176). For example, when parents act in line with 'future presents,' they recognize their responsibility for unforeseeable outcomes. That means, in a nutshell: Acting in line with 'present futures' entails a gesture of control; acting in line with 'future presents' entails a gesture of care. However, is there really no gesture of control involved, when caring for unknowable futures?

TSS is one field of research that has learned from the fatal history of linear control and has endorsed an understanding similar to Adam's notion of 'future presents.' Intergenerational justice and complex notions of risk and interdependence certainly are founding features of this field. Yet, despite a critical stance with regard to the socio-ecological consequences of control, the systemic notion of control that emerged in the 1960s lives on in TSS. One exemplary concept, as illustrated in Graph 1, holds that solutions to complex socio-ecological problems require three consecutive stages of knowledge generation (cf. Becker 2002; Grunwald 2007; Pohl and Hadorn 2007; Jahn, Bergmann, and Keil 2012; cf. Bornemann and Christen in this issue)

> [...] solution options should be composed of different types of knowledge. They should (1) be based on, at least, a sufficient understanding of the problem (descriptive-analytical/system knowledge); (2) be guided by a coherent and sustainability-inspired vision (normative/target knowledge); and (3) outline concrete transition and intervention strategies, i.e. action plans that detail how to resolve the problem and reach the vision (instructional/transformation knowledge). (Wiek and Lang 2016, 33)

According to this imaginary of collaborative research, intellectual development and the social negotiation with potential collaborators go hand in hand during the course of collectively deliberating and adapting to a complex set of goals (Van de Hel 2016). More currently, this process of launching a collaborative research project was exemplified by the concept of 'co-design,' which

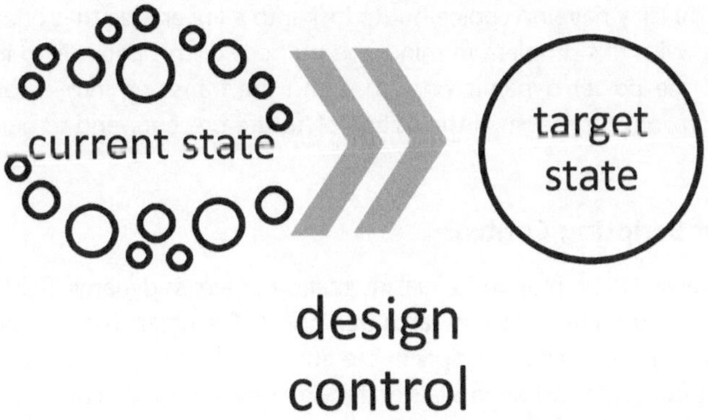

Graph 1. The concept of control in systems analysis.

describes the 'first phase of the knowledge co-production process, in which researchers and non-academic partners jointly develop a research project and define research questions that meet their collective interest and needs' (Moser 2016, 108). Co-design thus implies a social deliberation and an epistemic step-by-step progression. The given means and desired ends are thereby purposefully aligned, while somehow mastering the complexity of transdisciplinary mediation.

Along these lines, transdisciplinary sustainability methodologies describe collaborative research in a historical analogy with temporal control in systems analysis (depicted in Graph 1): In separating two different phases (the current state and the so-called target state), control represents the design process that takes a present endeavor to a desired future. Before a goal is reached, actors attempt to (re-)arrange the apparent means or to keep competing goals in check (cf. Hoos 1972; Jardini 2000; Wildavsky 1966). Even in the current concept of 'co-design,' where design is supposedly democratized, the underlying concept of control remains intact. It is ironic that the 'industrial hubris' of seeking to control time (Adam 2004, 147), as exemplified by the methods of systems analysis, continues to shape research practices in critical areas such as TSS.

Recently, transdisciplinary sustainability researchers reactivated a processual and relational understanding that collaborators 'learn to collaborate while collaborating' (see also Klein 1996; Vilsmaier, Brandner, and Engbers 2017; Van Breda and Swilling 2019). This implies a critique of design control. By contrast to 'planning forwards, towards or backwards from the future' van Breda and Swilling suggest that 'the present is not a burning platform between past and future, but rather where both meet in contested uncertainties expressed in both a multiplicity of experimentations and processes aimed at building up future imaginaries.' (Van Breda and Swilling 2019, 831). Social practices are seen as stretching over time and space so that the future represents an actualization of latent potentials (cf. Adam and Groves 2007). One consequence is a strong ethos of care, acknowledging that future harms unfold through present actions.

Both concepts of time – 'present futures' and 'future presents' – shape the predominant frameworks of epistemic control in TSS. A discussion on this ambivalence is vital for TSS, because this field in particular is set out to uncover and transform, by means of collaborative research, the hubris of industrial governing. However, in turn, the challenge of transforming control through collaboration begs the question: Which form of control is an appropriate alternative for collaborative research projects to realize a transformative agenda? In the following, I go beyond the affirmative notion of care, and seek to carve out a form of control that corresponds to the standpoint of 'future presents.' However, based on my concept of scripting control, I also argue that the capacity to reflect on temporal control is itself socially preconfigured. The capacity to learn to collaborate, the space for reflection, and the legitimacy of control are shaped early on in the course of the power dynamics of collaborative research. While the practical feasibility and appropriateness of linear control may be questioned by collaborating researchers, it is not unthinkable – but rather likely – that a corresponding managerial practice would emerge anyway.

Latecomers in particular may have no choice but to lock-into a linear idea that orients all tasks towards a desired target state. With this problem in mind, the purpose of this paper is to identify a concept of control that captures the power dynamic prior to – and the transition into – collaborative research. I propose the following concept to match the notion of 'future presents' and to guide the analysis.

2. The Concept of Scripting Control

The early sociologist Lewin (1982) mapped social interaction orders as dynamic fields of mutual expectations and desires. Following on from this research tradition, STS scholars have introduced the notion of 'scripts' (cf. Mattozzi 1987). I understand scripts in the ethnomethodological sense as 'constitutive rules' (Garfinkel 1963). Be it through social role models or through physical architecture – scripts exist as socially and materially configured drafts of action and, in their co-presence with other drafts, determine which options for action are considered plausible, legitimate, and appropriate. Scripts are not one-dimensional, but involve worldviews that shape organizational–technological learning processes (Vaughan 1996, 197), they entail design processes that incorporate certain logics into technologies (Akrich 1992), or they can be represented by technologically-predominated repertoires of user behavior (Akrich and Latour 1992). In constantly rearranging and countering certain constrictions, scripts facilitate how actors make sense of a complex sociospatial setting; they open-up and narrow-down the one's room for maneuver.

I want to emphasize the temporal and the power dimension of scripts. Essentially, scripts are a device of control, but simultaneously contradict a hierarchical understanding. Or, as Oliver Escobar wrote: 'Scripting is [...] not necessarily something that the "powerful" impose on the "powerless", but a transformative practice at play' (Escobar 2015, 16). This approach can be operationalized in a process-sensitive manner in reference to auxiliary verbs such as 'vouloir' (want), 'pouvoir' (can), and 'devoir' (should) (cf. Mattozzi 1987). Thus, for example, research grants establish a whole series of:

- Incentives and desires (want),
- Competences and opportunities (can), and
- Permissions and constraints (should).

One important specificity of scripts as a device of control is their radical embeddedness. As Suchman (1987, 295) writes: 'Emerging as endogenous resources for ordering "from within", plans, scripts and other ordering devices are woven intricately into the fabric of everyday activity.' Thus, scripts emerge from within the controlled processes, while representing embedded attempts at control. Consequently, there is interaction between several scripts over time, which shapes how reflexive or emancipatory a collaborative research can be with regard to complex control regimes. Moreover, scripts are temporal

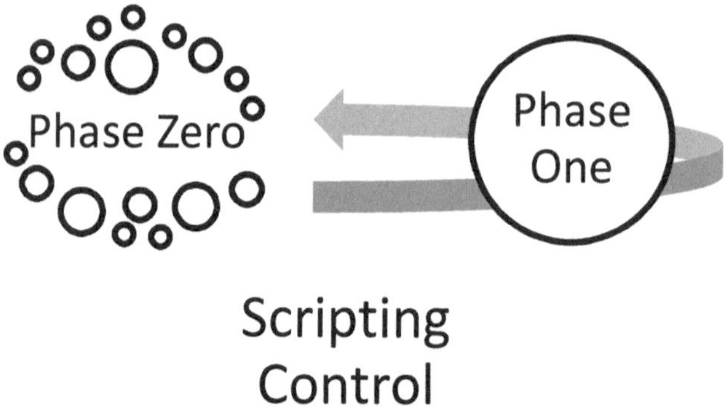

Graph 2. The concept of scripting control.

and spatial interfaces. A script is 'to be enacted, and it evolves when performed. The script is thus a heuristic to analyze the political assemblage that joins backstage and frontstage [...]' (Escobar 2015, 16).

Accordingly, the concept is particularly suited to avoiding two fallacies: Firstly, a conventional environmental determinism is evaded, so that actors are not seen as passive objects of design. Secondly, one sidesteps the risk of endorsing an actor-oriented voluntarism in which actors would be seen as entrepreneurs of social control. Especially in the temporal perspective it becomes clear, by contrast, that actors seek to co-determine what a project may become, without being able to assume linear control, let alone the ability to 'backcast' a pathway to get there. One's standpoint in time, as illustrated in Graph 2, directly corresponds with a decentralized 'locus of control' [...]. 'When the future is no longer thought to pre-exist but is approached instead as a realm to be shaped by human will, then potential outcomes are continuously shifting and changing' (Adam and Groves 2007, 164).

For my analytical purposes with regard to control in collaborative research, scripts serve as a tool to reconstruct the transition from backstage to frontstage, which is best expressed by the gerund 'scripting.' Scripting control, as I would term it, implies that control, for instance the quarrel to define a project leader, is in itself a bone of contention. Simultaneously, however, the early moments of an endeavor also entail the active sense of controlling, for instance when a limited set of collaborators defines who else to invite. Scripting control is thus a concept of practice that corresponds with the notion of 'future presents,' but contrasts with 'present futures' (cf. Adam and Groves 2007): Project initiators cannot simply backcast or forecast their project and a sense of control about phase zero is unlikely. Yet, these formative moments in a project offer actors ample opportunities to enroll into, to enrich, or to topple an interwoven set of scripts. Any script thus functions as a counter-script to others. Acting along the lines of 'future presents' therefore is not only a practice of care as Adam and Groves emphasize (2007); it is a relational struggle for control.

I use the two concepts of control – design control and scripting control – as ideal types to fit the two temporal concepts and to empirically describe different modes of control in collaborative research. This approach requires a temporally sensitive research design described in the following.

2.1. Methodology and Data Construction

Abbott (1995), among many others, has pointed to reconstructive approaches that trace sequences of social interaction. In a social arena, the interaction of several groups and structures can be followed over time, while the social arena is not seen as a container but as an emerging possibility space. The essential cornerstones of this sequential research style are the observation of interaction processes in the course of empirical time, especially through participatory observations, and the intimate knowledge of the action context of all groups involved.

While I was already part of the project presented in this case study, I drew on a complete survey of all internal discussion protocols that the initiators had collected during the build-up phase from 2008 to 2012, which were made available to me in electronic form. In consultation with an initiator of the project (and the project management), I added his notes from that construction phase. I also draw on 29 guideline-oriented interviews with sponsors and co-initiators (eleven interviews), volunteers (nine interviews), and permanent employees (seven interview partners, some interviewed several times), who shed light on the project from various angles. The empirical data are completed by 18 participating observations in several organizational parts and approximately 20 transdisciplinary interventions in the form of conversations with the two project leaders, four reflection sessions, and six meetings with a task force within the framework of an internal training process.

The data analysis consists of a retrospective sequencing of the accumulated and chronologically ordered data. Thus, singular events are examined and arranged so that a visualized and multi-layered sequence emerges as the object of interpretative analysis. This procedure makes the origins, conditions, and consequences of the scripts explicit, which can then be checked for their interplay in the course of a subsequent situation.

3. Phase Zero: Reconstructing the Interaction of Scripts

In 2012, I was commissioned to write a doctoral thesis on regional education as part of a third-party grant titled 'innovation incubator' at Leuphana University Lüneburg. The research program aspired to regional economic growth and included various interdisciplinary and qualification projects that all were expected to collaborate with regional partners from the economy and civil society. A more precise, no less economic, framework of interpretation was raised by another source: 'Jeremias Herberg's PhD project is concerned with the shortage of skilled workers.' With this press announcement, the educational organization that I worked with wanted to make public the collaborative relationship that had from the outset been tied to the third-party-funded doctoral fellowship. As part of the two-year collaboration that followed, I visited the organization about once a month and the research topic developed far beyond the scope of economic interpretation. Nevertheless, this initial moment makes clear that the present work was not only financed by, but also socially embedded in, the framing of a regional economic problem. This poses a problem of reflection: How to rearrange the economic research context in line with an open-ended and academically grounded PhD project?

Very indirectly, through a long history of interwoven scripts, my PhD position goes back to an open day held at a local grammar school in 2008. The newly elected Christian Democrat (CDU) constituency member of the state parliament met the school principal – according to his own statement, a 'particularly entrepreneurial representative of his guild' (Interview 2). The latter reported in the interview:

> And as it is appropriate, the school principal guides special guests through the house. [...] He and I then went through natural sciences, amongst other things. And I had the idea and asked him: What do you think about setting a scientific focus and perhaps looking for sponsors for the natural sciences rooms' equipment, that they consider funding one or two things in addition, which are not intended for a regular school building. He liked the idea. (Interview 1)

In the above quote it becomes clear that the headmaster wants to give his school a scientific profile by using a situational opportunity as well as his personal network in politics and among private sponsors. However, the idea failed soon after the open day. The administrative district did not allow for one public school to be better equipped than neighboring schools, much less be funded by private persons or businesses.

After this rejection, step by step, an alternative plan unfolded that placed a new educational organization in-between education and regional industries. Instead of back-casting the project against a set goal, the school principal and a growing group of allies embarked on a long journey of social networking and conceptual modification. They eventually put forward the idea that extracurricular learning opportunities would enable students of all school classes and school types to engage in a creative form of scientific–technical learning that directly promotes skills oriented toward employment. In organizational terms, many participants still argue today that the project's 'unique selling point' is that it is not supported by public or private sector organizations, but is instead a cross-sector organization and in that sense relatively independent of any one-sided control (interviews 1–3). A supporter of this idea expresses very aptly the intermediate position implied: 'We want to be an extracurricular place of learning, not another school. [...] But neither are we a company. We are somehow in-between' (Interview 3). In that sense, the project initiators essentially tried to control for the project's intermediary position. The corresponding challenge for the initiators was to fend off unilateral attempts at appropriation. This was achieved through interweaving multiple scripts and through a process of 'frame-shifting' (Cornelissen and Werner 2014). The following account illustrates this adaptive and multi-referential strategy. I will argue that it stands in a stark contrast to the ideal of 'co-design'.

3.1 From Educational 'Shoulds' . . .

The concept of extracurricular learning did not stem solely from the fact that in 2008 it had failed as a school project. According to the interview reports and some hints in internal minutes of the talks, the same frame had also appealed to the Prime Minister of Lower-Saxony and to the Ministry of Science and Culture (MSC), which already promoted similar learning locations. The early days of the project were thus already shaped by multiple scripts. Representatives of several ministries attended a project meeting hosted at the Ministry of Education (ME), so that the initiators could mobilize and use mutual audience roles. During the discussion, the state chancellery suggested the prospect of in-house subsidies, the ME offered new contacts to similar projects, and the MSC spoke of possible connections between the proposed project and universities (interview 2, internal correspondence). In short, the project emerged with a situated combination of education policy and science policy scripts. More specifically, a certain coercive pressure from the educational side ('should') was combined with curiosity from a science policy side ('want'). In mobilizing these scripts, it almost seems as if in the project initiators' eyes the future of their project represents an actualization of latent potentials (cf. Adam and Groves 2007).

The scripts they mobilized were rather fragile and non-binding. The ME soon withdrew its earlier commitments, and the MSC committed only to facilitate exchange between extracurricular institutions. In contrast to the MSC's position in this case, an earlier meeting in Hanover expressly supported collaboration between schools and universities through extracurricular education. This strategic moment led to another opportunistic 'frame-shifting' (Cornelissen and Werner 2014) in the project: There was a lack of science locations in the district of Harburg; more specifically, there were locations, but no civil society involvement, as in Lüneburg. Three years later, the CDU member of parliament, who had initially taken an interest in the school principal's idea, reinterpreted the effort as follows:

> Sure, we have the universities in Hamburg, but that the university is as socio-politically integrated as in Lüneburg and that there is a sense of togetherness in university and science within the district of Harburg – that cannot readily be claimed. [...] Therefore, the challenge we have set ourselves is to network business, education, and science, if we succeed, with great people [...]. And perhaps you will make the first step [directed to the author in reference to the doctoral position]. (Interview 2)

It becomes clear that in the interstice between education and business, lofty expectations and diverse framework conditions come together, but they also represent a gateway for power dynamics; for the attribution of role and performance expectations, but also for the acquisition of political attention and financial resources. The project initiators willingly incorporated a multiplicity of scripts and, for the moment, accepted the multiple identities that they were assigned. Or, as one interviewee puts it: 'You are [like] a school and/or a company, you serve the schools and/or the companies.' The interaction of scripts thus defines 'a framework of action together with the actors and the space in which they are supposed to act' (Akrich 1992, 208). The initiators thus endorsed a highly relational and almost opportunistic approach. But did they thereby abstain from a gesture of control?

3.2 ... to Science-Policy 'Wants'

Following the initial lack of official support for the idea of a school-centered project, the initiators relied on the expansion and connection of fragmented links to non-educational fields. This also addressed the involvement of companies and economic policy actors, as explained by an administrative official:

> Because the economy – you have to differentiate what this is, of course, the 'economy,' so there are certainly more than a few companies in the region, but also the constituted organizations that deal with companies [German: Chamber of Crafts]. These and others say that there is already a need for junior staff and in the foreseeable future this will probably increase, especially in fields that can be assigned to these subjects. And that is why it is obvious to get these stakeholders to join us [...]. So it is not only in the sense of basic research in physics and chemistry that young people should be brought in as young researchers, but also with an eye to the fact that we have companies in the region that need people today and tomorrow who are interested in scientific and technical correlations. (Interview 9)

According to my interview partners, regional firms were facing a problem that in German administrative jargon is coined 'Mittelabflussprobleme' [German: problems in the outflow of funds]. That is, they were looking for legitimate recipients for available funding, but they needed a publicly justifiable reason and also wanted to avoid a 'sponsoring competition' (Interview 8) among charitable projects (Interviews 6, 7). The initiators therefore looked for a controlled way, for a 'unique selling point,' to make the project appear useful in the eyes of others. The underlying understanding of control may not resemble a linear strategy of design, but the initiators were certainly eager to influence and – more importantly – to strategically leverage the reputation of the project. In order to make the project attractive to funders, the educational and scientific interests that were previously generated were now to be translated into economic interests.

3.3 ... to Regional Economic 'Shoulds'

In the following months, the notion of a skills gap could be presented as being of key relevance to economic and educational partnerships (broadly defined as an imbalance between the skills taught in the education system versus those required for employment). A concerted strategic effort to consolidate this motif took place in summer 2009. The project initiators invited a broad range of prospective partners to a so-called sponsor meeting. The keywords that I take from the preparatory notes are clearly based on employer-driven patterns of interpretation, and on correspondingly framed needs. At the local level, the challenges faced by employers were taken as a starting point in order to respond to the 'unspoken question' (protocol sponsor meeting); that is, to give them valid reasons to financially support the project. At this sponsor meeting, the initiators showed themselves to be a 'presentation platform for companies' and deliberately distanced the project from schools; this time in favor of the institutional identity of the invited companies, which was welcomed in order to give the project a non-governmental color. In other words, the initiators communicated a controlled distance from public education in order to build a useful proximity to regional employers. Even an interviewee, who later became the chairman of the project's booster club, saw it that way in an interview:

> The public sector is not perceived a major actor. And it is quite good that the [name of the organization] did not come directly from the district, from the school. [...] There's always a certain amount of district administration and public authorities to it, but there are already companies being engaged. 'That's not a school; it's a place of extracurricular learning': This reputation is good for the project. (Interview 3)

In this quote, for the third time, yet another 'frame-shifting' (Cornelissen and Werner 2014) becomes clear: from a location of extracurricular learning to a science location to a relevant site of business development. The initiators successfully controlled for the intermediary position of their project by capitalizing on cross-sectoral audience roles. In particular, the notion of a skills gap gained traction as a multi-referential frame: on the one hand, to give schools and education policymakers the prospect of an economically relevant role; on the other hand, to create corridors for the public sponsorship and investment by companies. This strategy is only indirectly related to the eventual concept of the project. Yet, it is a purposeful way to expose the endeavor to the power-play between economic and educational framings. Scripting control thus occurs when collaborators and their desires, capabilities and duties are used as '[...] endogenous resources for ordering "from within" [...]' (Suchman 1987, 295).

A few months after the 'sponsor meeting' in December 2009, the MSC rejected the subsidy proposal. It was argued that student laboratories (the terminology used by the MSC) are currently supported by third parties, i.e. by universities and companies not by the state. Two months later the Federal Ministry of Education and Research (BMBF) withdrew the prospect of funding, which had previously been acquired via CDU networks in Berlin. To make matters worse, private sponsorship of extracurricular Science-Technology-Engineering-Maths (STEM) projects would preclude public funding – a project could seemingly not benefit from both. The intermediary position that the project initiators were initially able to focus on now turned out to be rather precarious and contradictory. This shows the dynamic fragility of phase zero. The project initiators, who pushed for the stabilization of earlier scripts, relentlessly relied on the support of the next partner – only to mobilize new scripts and to stumble into new relationships.

How did the actors consolidate phase zero, possibly relaxing its fast pace? The characteristic strategy, which can be described as a trading of scripts, cumulated in winter 2009 when the group of initiators organized a workshop. It was at this public event that the project first made itself visible, and through which public (i.e. cross-field) legitimation of the project was to be strengthened. At the event, a two-step strategy can be discerned in which: (1) the promotion of regional science became a legitimate basis for (2) the acquisition of EU economic development funds. The coercive script from one field was thus related to a capacity-related script from another field. More precisely: During the workshop, the CDU member of the state parliament built bridges from science policy commitments to the economic framing of the extracurricular educational site to the involved operating costs. He argued that if the project could be presented as beneficial for the regional economy and linked to similarly framed university research, then it might be possible to secure EU funding to make up for the previous throwbacks. This economic narrative also convinced the educational and public policy audiences in attendance, so that their mutual relationship through the emerging organization was legitimized on the basis of the skills gap as a 'storified script' (cf. Escobar 2015). Already during the workshop, this strategic reasoning triggered the willingness of the various collaborators to formalize the project by discussing mutual obligations on the basis of a written document. This document, which was created through an unconsolidated strategy, would represent the basis for a corresponding grant proposal. This moment shows how the 'phase zero' of the project had to be pushed through in a rushed manner, until the patchwork of 'wants', 'shoulds' and 'cans' could finally be reconciled in a more solid framework. Soon after, the project managed to benefit from a large EU research and regional funding program that the Leuphana University of Lüneburg was awarded with the help of the MSC. This is one of the origins of the present doctoral project.

4. Limits of Scripting Control and the Transition to Phase One

The case study shows that the decoupling of the collaborative project from the school context through several steps led to a more diverse and financially strategic yet vulnerable affiliation. In the course of building the project, the initiators needed to find collaborators before securing funding, while at the same time the funding challenge formally tied them to certain collaborators. In that sense, diverse groups and concerns were fruitful during phase zero as a strategic opportunity structure. The organization, however, did not act as a purposeful broker, but rather as a multi-referential bone of contention. The initiators were not entrepreneurs of social control or designers of a visionary idea, but willful opportunists adapting to constantly emerging scripts.

The described phase zero is characterized by a highly decentralized 'locus of control' in both temporal and social terms (Adam and Groves 2007, 164). However, by contrast to Adam and Groves' affirmative understanding of 'present futures,' the initiators followed the relational notion of time with a strong will to dominate the project's future based on their political capital. 'Scripting control' thus is a relational, yet forceful way to shape the beginning of collaboration. Although scripting is not hierarchical but transformative (Escobar 2015, 16), the process of scripting control transformed the project into a hierarchical organization. How did this transition come about and what does it mean for temporal concepts of control? In the following I argue that the limits to 'scripting control' also define the temporal shift into phase one. I illustrate this argument based on my own role in the project under examination.

4.1. From Scripting Control to Design Control

When I was starting to conceptualize my perspective on the collaborative process, the group of project initiators already looked back on the described history. In other words: What was my phase zero was part of their phase one. In collaboration with volunteers and myself, an assignment of tasks, means, and goals had emerged from the regional economic positioning of the project so that the question now became how the daily work would meet the often lofty expectations. That connection, between research and learning on the one hand and economic scripts on the other, became clear in the above cited draft of a press release. Here, and over the following years, other stakeholders and I were confronted with consecutive regional economic framings. Yet, although my PhD research was originally scripted based on research grants and economic discourses, those scripts were not coercive. I was able, in agreement with the project management, to launch a counter-script. I reformulated the press release and instead described the project as a learning organization – a characterization that I assumed at the time would re-open the multiplicity of scripts, thus granting more conceptual openness and cross-sector learning (see Box 1).

Box 1. Excerpt from press release (October 2012).

Doctoral Research Progresses
Jeremias Herberg's doctoral thesis focuses on two questions: How does this extracurricular place of learning develop and how does it cooperate with its environment? He coined the phrase 'the [name of the organization] is a learning organization.' A nice thought: we do not just want to teach others something, but we and the organization also learn during this process.

The project's phase zero as presented above supports the underlying intuition: When one script is problematic to accept, it may be rearranged, combined, or undermined by emphasizing the importance of another one.

At first, my framing was taken up and shortly after the press release a conversation took place about how the organization could actually learn. The two managers of the organization and I wanted to create a platform in which the leadership would have the opportunity to reveal and reflect its own way of thinking, independent of the short-term expectations of other collaborators. In a way, the aim was to decelerate the dynamic of scripting described above. A total of four such sessions took place, preceded by one conceptual meeting (February–August 2013). Thematically, one of the managers took up what she perceived to be a 'winged word.' The 'learning organization' framing, which I had introduced in order to give the conversations a larger, more reflexive context, was an appealing notion from her managerial perspective. Consequently, it was her request that encouraged me to translate the initially spontaneous reaction to the preformulated press release into a comprehensive concept proposal, which I soon submitted. In the organizational learning sessions, which I consequently initiated together with the two managers, they lamented the lack of focus among the predominantly older and male group of volunteers. Against this background the process of a 'learning organization' was welcomed due to its gentle overtone. Yet, my proposal was also accepted because of the managerial intention to streamline the educational program. The point of contention was no longer the acquisition of grants as in phase zero. The managers now sought to define the pedagogic skills and strategic orientations of the volunteers that had joined the project in the meantime.

In the first meeting, the challenge that the manager described is that there is a lack of collective awareness among the many volunteers that contributed on the basis of their personally held motivations. To complicate the matter, in the course of the operative start of the project, the incorporation of volunteers and the narrative of a shortage of skilled workers also suggested increasing pressure to consolidate task divisions and to achieve results. In dealing with this new pragmatic pressure, I suggested that the term 'learning organization,' is helpful to prolong the conceptual openness that had characterized the organization initially. This proposal, which would also give me space to design my involvement before it would become part of the collaborative dynamic, did not work out as anticipated. The term was subsequently appropriated by the organization's leadership in order to narrow-down the skills that were expected from volunteers and employees. In that sense, the notion was welcomed in as a kind of Trojan Horse: especially at an educational organization, the imperative of learning would have to be welcomed, while simultaneously promoting a set of restrictions that the organization's management could use to tame the complex set of expectations. This script, which I had launched myself, entailed certain wants, capacities, and possibly even coercions, which the managerial leadership was able to adopt in order to rationalize the collaboration among staff and partners.

In phase zero, relatively weak promises were turned into material resources. However, while my collaborators were moving on, I attempted to reopen the relatively focused project and develop a phase zero for my doctoral research. The term 'learning organization' that I suggested for this purpose implies wishful thinking about the need to open-up to heterogeneous ways of thinking. The failure to effectively implement this framing arguably marks a temporal transition between phase zero and phase one. Scripts – previously used in the sense of an opportunistic and relational notion of scripting control – subsequently hardened as resources to be used for a more directional form of control.

5. The 'Dromocracy' of Collaborative Research

Paul Virilio, a pioneering theoretician on time and control, suggested a conceptual conflict between democracy and 'dromocracy' (Virilio [1977] 2006); that is, the momentary synchronization of diverging concerns and interests is superseded by the rule of time. This rule of time is driven by political tactics, military technology, and economic growth but can rarely be harnessed in an integrative and goal-oriented manner. Virilio's global diagnosis, which pinpoints the failures and frictions of temporal control in times of technological disasters, is reflected in the nitty-gritty of the presented case study: After a relatively open-ended beginning, the initiators, who initially had a school project in mind, succeeded in generating a viable patchwork of funding and partnerships. They did so by endorsing an industry-focused legitimation. The script of the skills gap was storified in front of multiple audiences, while in effect it helped to prioritize the interests of employers over educators. My own involvement eventually necessitated counter-scripts to make room for open-ended involvement and an academic dissertation. However, the launched counter-script was quickly appropriated, itself becoming a resource for control within the now hierarchical organization. The increased complexity of the collaborative process is thus driven by highly interrelated and, in effect, by a continually failing and constantly accelerating set of control attempts.

How can collaborative researchers – be they in STS or TSS – come to terms with the 'dromocracy' of collaborative research? As illustrated by the historical example of systems analysis, even reflexive responses to linear control can be used to avoid questions of responsibility. More recently, the practical notion of 'co-design' represents a more ethically minded approach. While acknowledging the power sensitivity of new beginnings, proponents of 'co-design' seek to resolve the intractability of emergent control regimes; they seemingly want to halt the 'dromocracy' of collaborative research. The presented case study, however, frustrates the hope for a democratic beginning as cultivated in TSS and beyond. Phase zero, as I have coined it, is likely to remain a transitionary moment, hastily navigated by means of mutual desires, duties, and demands. Recent work that has introduced relational theories to the sustainability literature has begun to discuss the corresponding limits to control (Van Breda and Swilling 2019; Freeth and Caniglia 2019). However, while the concept of scripting control may make an analytical contribution to this line of thought, from a normative stance it does not imply a more desirable control regime. By contrast to the directional gesture of design, the relational strategy of scripting implies less transparency about one's regulatory intentions. Scripts even disappear when enacted, so that actors can at best offer counter-scripts; at worst, they become inseparable from their scripted paths. This dynamic can hardly be democratized because scripting – albeit less directional and more processual than design – involves a concealed form of influence. By means of this process, the most powerful actors succeed in bundling a given interplay of scripts in their own interest. They can determine what is perceived as a 'common language' on the frontstage, while in the same gesture retreating backstage. This process is both a process of acceleration and collaboration, so that the difficulty of pinpointing or turning around control regimes potentially increases over time. We may even speculate whether Virilio's dromological law also applies to collaborative research: namely that greater speed (i.e. tighter temporal constraints) increases the chances of gridlock.

Future research about collaborative research or even future methodologies of transdisciplinary collaboration may go beyond the synchronous focus on (co-)design and the normative evaluation of more or less democratic forms of collaboration. Thus levelling the birds-eye perspective and endorsing the relational viewpoint, one can study collaboration from within the process and over time. This does not undermine, but specify the ethical quest for collaborative research: How to conceptualize the role of researchers when taking into account myriad projected expectations? How to conceive of scripts and counter-scripts as a conflictual, yet deliberative process? How to shape a collaborative process when a shared goal or language may never emerge as a permanent and broadly accepted footing of a project? These are just some of the questions that future research about the 'dromocracy' of collaborative research may explore.

Disclosure statement

No potential conflict of interest was reported by the author.

ORCID

Jeremias Herberg http://orcid.org/0000-0003-4388-796X

References

Abbott, A. 1995. "Sequence Analysis: New Methods for Old Ideas." *Annual Review of Sociology* 21: 93–113. doi:10.1146/annurev.so.21.080195.000521.

Adam, B. 2004. *Time*. Cambridge: Polity.

Adam, B., and C. Groves. 2007. *Future Matters: Action, Knowledge, Ethics*. Leiden: Brill.

Akrich, M., and B. Latour. 1992. "A Summary of A Convenient Vocabulary for the Semiotics of Human and Nonhuman Assemblies." In *Shaping Technology/Building Society. Studies in Sociotechnical Change*, edited by W. E. Bijker and J. S. Law, 259–264. Cambridge, MA: MIT Press.

Akrich, M. 1992. "The De-Scription of Technical Objects." In *Shaping Technology/ Building Society: Studies in Sociotechnical Change*, edited by W. E. Bijker and J. Law, 205–224. Cambridge, MA: MIT Press.

Becker, E. 2002. "Transformations of Social and Ecological Issues into Transdisciplinary Research." *Knowledge for Sustainable Development: An Insight into the Encyclopedia of Life Support Systems* 3: 949–963.

Cornelissen, J. P., and M. D. Werner. 2014. "Putting Framing in Perspective: A Review of Framing and Frame Analysis across the Management and Organizational Literature." *The Academy of Management Annals* 8 (1): 181–235. doi:10.5465/19416520.2014.875669.

Escobar, O. 2015. "Scripting Deliberative Policy-Making: Dramaturgic Policy Analysis and Engagement Know-How'." *Journal of Comparative Policy Analysis: Research and Practice* 17 (3): 269–285. doi:10.1080/13876988.2014.946663.

Freeth, R., and G. Caniglia. 2019. "Learning to Collaborate while Collaborating: Advancing Interdisciplinary Sustainability Research." *Sustainability Science*: 1–15.

Garfinkel, H. 1963. "A Conception Of, and Experiments with "Trust" as A Condition of Stable Concerted Actions." In *Motivation and Social Interaction*, edited by O. J. Harvey, 187–238. New York, NY: Ronald Press.

Grunwald, A. 2007. "Working Towards Sustainable Development in the Face of Uncertainty and Incomplete Knowledge." *Journal of Environmental Policy & Planning* 9: 245–262. doi:10.1080/15239080701622774.

Hoos, I. 1972. "*System Analysis in Public Policy - A Critique.*" Berkeley: University of California Press.

Jahn, T., M. Bergmann, and F. Keil. 2012. "Transdisciplinarity: Between Mainstreaming and Marginalization." *Ecological Economics* 79: 1–10. doi:10.1016/j.ecolecon.2012.04.017.

Jardini, D. 2000. "Out of the Blue Yonder: The Transfer of Systems Thinking from the Pentagon to the Great Society, 1961–1965." In *Systems, Experts, and Computers: The Systems Approach in Management and Engineering, World War II and After*, edited by A. C. Hughes and T. Hughes, 311–358. Cambridge, MA: MIT Press.

Klein, J. T. 1996. *Crossing Boundaries: Knowledge, Disciplinarities, and Interdisciplinarities*. Charlottesville: University Press of Virginia.

Lewin, K. 1982. *Feldtheorie*. Bern: Huber.

Mattozzi, A. 1987. "Rewriting The Script. A Methodological Dialogue About The Concept Of "Script" and How To Account For The Mediating Role of Objects." Paper presented at a colloquium at the University of Twente in September 1987, University of Twente, Enschede, The Netherlands.

Moser, S. C. 2016. "Can Science on Transformation Transform Science? Lessons from Co-design." *Current Opinion in Environmental Sustainability* 20: 106–115. doi:10.1016/j.cosust.2016.10.007.

Pohl, C., and G. H. Hadorn. 2007. *Principles for Designing Transdisciplinary Research*. Munich: Oekom.

Suchman, L. A. 1987. *Plans and Situated Actions: The Problem of Human–Machine Communication*. Cambridge: Cambridge University Press.

Van Breda, J., and M. Swilling. 2019. "The Guiding Logics and Principles for Designing Emergent Transdisciplinary Research Processes: Learning Experiences and Reflections from a Transdisciplinary Urban Case Study in Enkanini Informal Settlement, South Africa." *Sustainability Science* 14 (3): 823–841. doi:10.1007/s11625-018-0606-x.

van der Hel, S. 2016. "New Science for Global Sustainability? The Institutionalisation of Knowledge Co-production in Future Earth." *Environmental Science & Policy* 61: 165–175. doi: 10.1016/j.envsci.2016.03.012.

Vaughan, D. 1996. *The Challenger Launch Decision: Risky Technology, Culture, and Deviance at NASA*. Chicago: University of Chicago Press.

Vilsmaier, U., V. Brandner, and M. Engbers. 2017. "Research In-between: The Constitutive Role of Cultural Differences in Transdisciplinarity." *Transdisciplinary Journal of Engineering & Science* 8: 169–179.

Virilio. [1977] 2006. *Speed and Politics*. Cambridge: MIT Press.

Wiek, A., and D. J. Lang. 2016. "Transformational Sustainability Research Methodology." In *Sustainability Science. An Introduction*, edited by H. Heinrichs, P. Martens, G. Michelsen, and A. Wiek, 31–41. Dordrecht: Springer.

Wildavsky, A. 1966. "The Political Economy of Efficiency. Cost-Benefit Analysis, Systems Analysis, and Program Budgeting." *Public Administration Review* 26 (4): 292–310. doi:10.2307/973301.

Index

Page numbers in *italic* refer to figures.